T0276470

Apoptosis: Advanced Topics

Apoptosis: Advanced Topics

Edited by **Max Cowen**

New York

Published by Callisto Reference,
106 Park Avenue, Suite 200,
New York, NY 10016, USA
www.callistoreference.com

Apoptosis: Advanced Topics
Edited by Max Cowen

International Standard Book Number: 978-1-63239-074-5 (Hardback)

Printed in the United States of America.

Contents

Preface VII

Chapter 1 **Apoptosis and Clearance of the Secretory Mammary Epithelium** 1
Jamie C. Stanford and Rebecca S. Cook

Chapter 2 **Apoptosis and Activation-Induced Cell Death** 29
Joaquín H. Patarroyo S. and Marlene I. Vargas V.

Chapter 3 **Neuronal Apoptosis in HIV-1-Associated Central Nervous Diseases and Neuropathic Pain** 51
Mona Desai, Ningjie Hu, Daniel Byrd and Qigui Yu

Chapter 4 **Translational Control in Tumour Progression and Drug Resistance** 73
Carmen Sanges, Nunzia Migliaccio, Paolo Arcari and Annalisa Lamberti

Chapter 5 **Apoptosis During Cellular Pattern Formation** 95
Masahiko Takemura and Takashi Adachi-Yamada

Chapter 6 **Extra-Telomeric Effects of Telomerase (hTERT) in Cell Death** 107
Gregory Lucien Bellot and Xueying Wang

Chapter 7 **Programmed Cell Death in T Cell Development** 125
Qian Nancy Hu and Troy A. Baldwin

Chapter 8 **Drug Resistance and Molecular Cancer Therapy: Apoptosis Versus Autophagy** 155
Rebecca T. Marquez, Bryan W. Tsao, Nicholas F. Faust and Liang Xu

Permissions

List of Contributors

Preface

This book covers a broad spectrum of Apoptosis-related scientific research. Apoptosis is a programmed cell death (PCD). Since its discovery more than a century and half ago, apoptosis has been implicated in many processes in living organisms. It will be a valuable source for medical practitioners, researchers and students for updating their knowledge on recent advances in this field.

This book is a comprehensive compilation of works of different researchers from varied parts of the world. It includes valuable experiences of the researchers with the sole objective of providing the readers (learners) with a proper knowledge of the concerned field. This book will be beneficial in evoking inspiration and enhancing the knowledge of the interested readers.

In the end, I would like to extend my heartiest thanks to the authors who worked with great determination on their chapters. I also appreciate the publisher's support in the course of the book. I would also like to deeply acknowledge my family who stood by me as a source of inspiration during the project.

Editor

Chapter 1

Apoptosis and Clearance of the Secretory Mammary Epithelium

Jamie C. Stanford and Rebecca S. Cook

Additional information is available at the end of the chapter

1. Introduction

The development of the mammary gland occurs in four distinct phases: embryogenesis, puberty, pregnancy, and a post-lactational phase involving profound levels of cell death and tissue remodeling. This post-lactational phase is termed post-lactational involution. During embryogenesis, a solid epithelial bud is generated in the embryonic ectoderm. As this bud continues to grow in cell number, the epithelial bud invaginates into the underlying mesenchyme forming the nascent mammary epithelium. The mammary epithelium grows as solid epithelial cords, lengthening distally and branching to form the rudimentary epithelial network. At puberty, ductal elongation continues in a proximal-to-distal direction, and side branches appear along the ducts. The side branches also lengthen distally, and continue to branch. This pattern of distal growth and branching fills the mouse mammary fat pad with an extensively branched epithelium by the end of puberty [6]. Similar to what is seen during embryonic mammary development and patterning, the mammary ducts developing during puberty originally appear in solid epithelial cords. Apoptosis canalizes the luminal space within the ducts, allowing a patent conduit for milk to traverse through the breast epithelium [1, 7]. Ultimately, the rodent mammary epithelium is comprised of a continuous, branching network leading from the nipple to primary ducts and smaller ductules that terminate in terminal end buds (TEBs), blunt ends or alveoli. The inner luminal cells are separated from the basement membrane by an outer myoepithelial layer. Myoepithelial cells secrete basement membrane components to which the epithelium attaches, and that physically separates the epithelium from the stromal compartment.

Many morphological similarities exist between the mouse mammary gland and the human breast, although some distinctions exist. In the human breast, the cluster of epithelial acini arising from a single terminal duct, referred to as the terminal duct lobular unit (TDLU), is thought to be the milk-producing unit of the mammary gland. Therefore, the post-pubertal human breast harbors cells capable of milk production even in the absence of pregnancy whereas the rodent mammary gland does not. However, profound expansion and differentiation of the TDLU population in the human breast is still required in order to render lactation successful.

This expansion of the alveolar epithelium during pregnancy occurs in response to both local and systemic factors that drive mammary alveolar proliferation. In rodents, the entire secretory epithelium of the mammary gland develops during the gestation period of approximately three weeks, signifying a rapid 10-fold increase in epithelial content of the mammary gland. The mammary gland produces colostrum and then milk upon partuition. However, once offspring are weaned, the milk-producing lobuloalveolar cells are no longer necessary. Rather than being maintained, these secretory cells undergo programmed cell death in an exquisitely controlled and rapid process, while leaving the ductal epithelium intact. Dying cells are cleared from the post-lactational mammary gland rapidly and without causing acute inflammation, removing up to 90% of the total mammary epithelial content within the time span of just one week in rodent models. This returns the mammary gland to an almost pre-pregnant state, so that the changes in the mammary gland that accompany pregnancy, lactation and involution may again occur with each successive pregnancy.

The process of involution is complex requiring two distinct phases, the initiation of extensive cell death to remove the milk-producing epithelial cells, followed by the controlled influx of macrophages and other immune cell types to breakdown extracellular matrix, remodel blood vessels, replenish the adipocyte population in the mammary fatpad, and to phagocytically remove dead cells, residual milk, and debris. This review will focus primarily on the events controlling cell death that occur within the first days of post-lactational involution.

2. Body

2.1. Signaling mechanisms that control post-lactational apoptosis in the breast

In recent years, molecular regulation of post-lactational involution has been studied primarily in the mouse mammary gland, due in large part to the relatively rapid gestation and nursing period in mice, and to the extensive use of genetically engineered mouse models. These models, coupled with advances in transcriptional profiling have provided a detailed analysis of the dynamic cellular and molecular events occurring during the earliest days of involution, when the majority of programmed cell death occurs.

2.2. Milk stasis

Using teat-sealing to block milk delivery in a single mouse mammary gland, investigators demonstrated that a complex multi-step process initiating massive epithelial apoptosis is triggered by local stimuli produced in the sealed mammary gland, but not by changing levels of circulating hormones that are available to the remaining nine mouse mammary glands [8, 9]. These studies revealed that milk stasis is a primary trigger of post-lactational cell death in the mammary gland [8]. Accumulation of milk within the secretory luminal space might initiate cell death by causing a mechanical stretch of these cells, or of cell-cell junctions [10]. It is clear that mechanical stress, including cell stretching, can initiate biological responses in several epithelial and endothelial cell types, and may activate signaling pathways known to trigger cell death in the post-lactational mammary gland. For example, cell stretching induces STAT3 phosphorylation, inhibition of the survival factor AKT, and expression of Leukemia Inhibitory Factor (LIF), each of which are critical during early post-lactational involution for the induction of cell death, as discussed below. Another potential explanation of milk stasis-induced cell death is that accumulation of milk components, such as calcium, may trigger cell death [11]. In support of this hypothesis, transcription of the plasma membrane protein calcium-ATPase 2 (PMCA2), which transports 60–70% of milk calcium [12], is dramatically and rapidly reduced during involution, perhaps due to self-limiting negative feedback in an effort to control potentially toxic divalent cation levels [13]. Loss of the gene encoding PMCA2 (*Atp2b2)* in mice caused precocious alveolar cell death at lactation. Interestingly, PMCA2 expression is also regulated by enforced shape changes in mammary epithelial cells [13]. Stanniocalcin-1 (STC-1), a newly discovered mammalian hormone that accumulates nearly 3-fold upon milk stasis [14], has recently been implicated as an inducer of post-lactational involution [15].

2.3. STAT3

Transcriptional profiling studies of the mouse mammary gland at specific time points during lactation and post-lactational involution demonstrated that a specific subset of genes is dramatically induced within 12 hours of pup withdrawal, presumably in response to milk accumulation [4, 16]. It was hypothesized that this gene subset may represent potential 'master regulators' of programmed cell death in the post-lactational mammary gland. This idea has been largely confirmed using genetically engineered mouse models that disrupt key expression events, resulting in a delay in post-lactational programmed cell death.

The transcription factor Signal Transducer and Activator of Transcription (STAT) 3 was conditionally deleted in the mammary epithelium of genetically engineered mice, revealing its critical role in initiating the earliest events in post-lactational apoptosis [17-19]. While it has been known for some time that STAT3 regulates the expression of pro-inflammatory genes involved in the acute phase response (the early inflammatory response to tissue injury) [20, 21], and that many inflammation-related genes are expressed during post-lactational involution [4, 22, 23], these studies were the first to demonstrate the molecular similarities that exist between the involuting mammary gland and the traditional wound healing scenario [23], despite the fact that involution-induced transcriptional responses are directed primarily by epithelial cells, while wound healing-induced transcriptional responses are directed by cells of the immune system.

STAT3 is widely expressed, and is activated by tyrosine phosphorylation in response to numerous cytokines and growth factors [e.g. interleukin-6 (IL-6), IL-10, IL-17, IL-23, EGF] and tyrosine kinases (c-Src, Met, ErbB-2) [24]. Tyrosine phosphorylation of STAT3 allows

STAT3 dimers to translocate from the cytoplasm to the nucleus, where STAT3 binds to sequence-specific DNA elements in the promoters of STAT3 target genes. STAT3 activates gene transcription of many inflammation related genes, and can also repress the transcription of others. Although STAT3 transcriptional activity is associated with cell survival in several cell types including lymphomas and solid tumor epithelial cells [25-28], STAT3 takes on a different role in the post-lactational mammary gland, where STAT3 is required to initiate cell death. In the absence of STAT3, cell death was abrogated for at least 6 days after pup withdrawal, despite milk stasis [18, 19]. Conversely, loss of suppressor of cytokine signaling (SOCS)-3, a negative regulator of STAT3, accelerated involution by increasing the rate of cell death following pup withdrawal [29, 30].

A number of genes regulated by STAT3, such as CAAT/enhancer binding protein (C/ebp)δ, oncostatin M (OSM), OSM receptor (OSMR), and insulin-like growth factor (IGF) binding protein (IGFBP)-5, are also required in the post-lactational mammary epithelium to initiate cell death [31, 32]. OSM, a cytokine normally produced by macrophages but in this case produced by the mammary epithelium, is required during post-lactational involution, since OSMR knockout mice exhibited delayed involution [33]. Loss of C/ebpδ, a transcription factor involved in the acute phase response, also delayed mammary gland involution [31]. Because STAT3, and many target genes activated by STAT3, are critical triggers of cell death during post-lactational involution, it is likely that STAT3 lies at the apex of a transcriptionally-activated signaling cascade that is required to initiate cell death in the post-lactational mammary epithelium. This role of STAT3 as an apoptosis inducer lies in contrast with observations that STAT3 is frequently activated in several cancer entities [28], correlating with heightened malignancy [34, 35]. Further, constitutive STAT3 activity promotes tumor formation in skin [36] and lung [37, 38]. The apparent discrepancy may be related to tissue specificity of STAT3 activity, or the activity of STAT3 in the tumor microenvironment (for example, in inflammatory cells) versus its role in the epithelial compartment of the tumor.

2.4. NF-κB

Although the role of STAT3 in the induction of post-lactational apoptosis is clear, STAT3 alone is insufficient to induce involution in the absence of nuclear factor-κB (NF-κB) signaling [39]. NF-κB comprises a family of five structurally and functionally related transcription factors [40]. Based on their transactivation properties, NF-κB proteins are divided into two classes: Class I consists of RelA/p65, RelB, and c-Rel, while Class II includes NF-κB1/p50 and NF-κB2/p52. Each can dimerize in almost any combination but only class I proteins possess the C-terminal transactivation domains required for NF-κB-mediated transcription of target genes. Under basal conditions, NF-κB dimers are sequestered in the cytoplasm bound to the protein Inhibitor of κB (IκB). Several signaling pathways can activate the IκB kinases (IKKs) that phosphorylate IκB, thus liberating NF-κB dimers and allowing their nuclear translocation, where they bind to specific DNA sequences in target genes.

Among this family of transcription factors, two NF-κB subunits, RelA (p65) and p50 are expressed at different levels in the mammary epithelium throughout mammary gland development. Furthermore, NF-κB activity as measured *in vivo* using a transgenic NF-kB reporter model demonstrated two major peaks of NF-κB-mediated trans-activation: one that occurs during pregnancy, and another that occurs during involution [41]. These data are consistent with reports showing that NF-κB induced transcription of pro-survival genes [42-45], and other reports showing that NFκB activated transcription of pro-apoptotic genes [46-48]. Therefore, it is possible that NF-κB might drive cell growth and survival in the breast epithelium in some cases, but may regulate breast epithelial cell death in others. In support of the idea that NF-κB might promote cell death, increased NF-κB activity is rapidly induced after weaning, with strong increases seen within one hour of milk stasis in mouse mammary glands. NF-κB activity remains elevated through the first four days of murine post-lactational involution. Furthermore, loss of NF-κB signaling in a genetically engineered mouse model of conditional IKK-β disruption decreased post-lactational NF-κB signaling, resulting in decreased caspase-3 cleavage and delayed post-lactational apoptosis [39], confirming the importance of NFκB signaling in post-lactational cell death of the secretory epithelium. Conversely, constitutively active IKK-β increased NF-κB signaling, thus causing accelerated induction and higher rates of apoptosis during post-lactational involution [49]. Even in the absence of milk stasis, constitutively active IKK-β was capable of inducing apoptosis in the mouse secretory mammary epithelium, and therefore interfered with successful lactation by nursing dams.

2.5. Akt/PI3K

Intense interest in survival signaling pathways has revealed that phosphatidyl inositol 3-kinase (PI3K) is a potent regulator of cell survival [50, 51]. Cancer cells frequently utilize PI3K signaling to promote cell survival under conditions of hypoxia, nutrient stress, or even to escape the cytotoxic effects of therapeutic anti-cancer treatments. It is clear, however, that non-transformed cells also use the PI3K signaling pathway to promote cell survival, and that increased PI3K signaling can interfere with physiological cell death [52]. PI3K is a heterodimer comprised of p110 (the catalytic domain) and p85 (the regulatory domain) [50, 53]. Under basal conditions, p85 represses the catalytic activity of p110. However, SH2 domains in p85 interact with phosphorylated tyrosines within YxxM motifs of receptor tyrosine kinases (RTKs) such as the insulin-like growth factor (IGF)-1 receptor (IGFR) or adaptor proteins, such as the insulin receptor substrate proteins. This relieves p85-mediated inhibition of p110, allowing p110 to phosphorylate phosphatidyl inositol 2-phosphate (PIP2), thus generating PIP3, a powerful membrane-associated second messenger that recruits pleckstrin homology (PH)-domain containing proteins to the cell membrane. PDK1 and Akt are two PH-domain containing proteins recruited to the membrane in response to RTK activation [52, 54, 55]. PDK1 is a serine-threonine kinase that phosphorylates Akt, another serine-threonine kinase that stimulates cell survival by interacting with members of the Bcl2 family of apoptosis regulators [56-58], which are also involved in the induction of cell death during involution [59-61]. For example, mammary-specific loss of the Bcl2 family member Bax, a known cell death inducer, decreased apoptosis during early involution [60, 62-64], while overexpression of Bax within the secretory mammary epithelium increased post-lactational apoptosis and promoted precocious STAT3 activity [7, 65]. Loss of the anti-apoptotic protein Bcl-xL in the mammary gland during post-lactational involution accelerated cell death [66, 67]. By inactivating Bax and activating Bcl-xL, Akt activity increases cell survival in the secretory mammary epithelium.

The role of PI3K/Akt signaling in suppressing post-lactational apoptosis is supported by genetically engineered mouse models that result in increased PI3K/Akt signaling. For example, a mouse model in which mammary-specific expression of myristoylated p110α [68], a modified p110α that is restricted to the cell membrane, resulted in aberrantly elevated PI3K activity in the mammary epithelium and delayed post-lactational involution. Similarly, mammary-specific transgenic overexpression of Akt1 or Akt2 promoted cell survival and delayed post-lactational involution in mice [69, 70]. Conversely, ablation of Akt1, but not the ablation of Akt2 or Akt3, promoted apoptosis and accelerates involution [71], demonstrating isoform-specificity in the gene-dosage effects of Akt (overexpression versus ablation), and highlights the importance of Akt1 in the post-lactational mammary gland. Other studies demonstrated that Akt signaling is sustained during lactation by prolactin signaling [72-74]. This observation was confirmed in an independent transgenic mouse model of mammary-specific STAT5 activation, in which STAT5 activity, when aberrantly sustained through post-lactational involution, upregulated Akt1 transcription and impaired apoptosis. These studies suggest that high levels of prolactin-induced STAT5 activity, as seen during lactation, maintains Akt1 expression and activity to promote cell survival, but when lactation ceases STAT5-induced Akt expression must be depleted in order for cells to undergo apoptosis [75].

Like prolactin signaling, other ligand-activated signaling cascades are capable of driving PI3K/Akt signaling during lactation, and if not turned off, can delay post-lactational apoptosis. For example, cell signaling initiated by IGF-1, which activates IGFR thus causing tyrosine phosphorylation of insulin receptor substrate proteins [76], potently activates the PI3K/Akt signal transduction cascade in the mammary epithelium during lactation. Overexpression of IGF-1 in the mouse mammary gland delayed post-lactational involution, suggesting that suppression of IGF-mediated cell survival is required for apoptosis to occur, and supporting the idea that PI3K signaling must be interrupted to initiate post-lactational apoptosis [77]. IGF-1 bio-availability is tightly controlled by IGF binding proteins (IGFBPs), which can sequester IGFs in the extracellular microenvironment of the mammary epithelium [78]. Consistent with the ability of IGF-1 to interfere with post-lactational cell death, one of the earliest transcriptional events during post-lactational involution is the upregulation of IGFBP-2 mRNA (4-fold), IGFBP-4 (6-fold) and IGFBP-5 mRNA (50-fold) [79]. This profound increase in IGFBP-5 is also seen at the protein level, and is conserved across several species. Increased expression of IGFBPs may limit IGF1-induced signaling, thus limiting IGF1-induced PI3K/Akt signaling [77, 80, 81]. Transgenic overexpression of IGFBP-5 in the mouse mammary gland increased caspase-3 cleavage (an indicator of apoptosis) and decreased the expression of the pro-survival factors Bcl-2 and Bcl-xL [80-84], suggesting that IGFBP-5 is pro-apoptotic. An IGF-1 analogue which binds weakly to IGFBP-5 partially overcame IGFBP-5-induced cell death during post-lactational involution, suggesting that IGFBP-5 was acting, at least in part, by inhibiting IGF action. Conversely, *Igfbp5* null mammary glands exhibit delayed post-lactational apoptosis [78].

2.6. TGFβ3

While prolactin, IGF-1, and several RTK-activating ligands can activate PI3K/Akt signaling to promote cell survival, other ligands are capable of inducing cell death during post-lactational involution, such as leukemia inhibitor factor (LIF) [85-87], serotonin [88], Fas ligand (FasL) [89] TRAIL [90], and transforming growth factor (TGF)-β3. Transcripts encoding TGF-β3, but not TGF-β1 or TGF-β2, substantially increase in the milk-producing cells during post-lactational involution [91, 92]. This rapid induction of TGF-β3 transcription in the secretory mammary epithelium occurs as early as 3 hours after pup withdrawal in response to milk stasis [93], and is among the most rapid gene expression changes occurring in response to post-lactational involution, suggesting that TGFβ3 might be an initiating signal for cell death during involution. It would be interesting to determine the impact of mechanical stress on expression from the TGFβ3 promoter. Consistent with the proposed role for elevated TGF-β3 in inducing apoptosis during involution, transgenic over-expression of TGF-β3 in the secretory cells of the mouse mammary gland accelerated apoptosis during early post-lactational involution. Conversely, loss of TGFβ3 reduced post-lactational apoptosis by nearly 70% [93], suggesting that autocrine TGF-β3 signaling initiates cell death following pup removal.

The importance of the TGFβ3-induced signaling pathway for post-lactational apoptosis has been further investigated in genetically engineered mouse models. For example, loss of the TGFβ-regulated transcription factor Smad3 decreased post-lactational apoptosis by nearly 40% [94]. Similarly, loss of TGF-β receptor type II (TβRII) in the mammary epithelium, or transgenic expression of dominant negative (DN) TβRII decreased apoptosis during early post-lactational involution [95-97], consistent with a critical role for TGF-β3 signaling through TβRII and Smad3 to induce apoptosis during early involution. However, there is some discrepancy regarding the role of TGFβ signaling during involution, as transgenic expression of constitutively active TβRI decreased apoptotic cells in the mammary gland [98]. Perhaps elevated TβRI signaling activates signaling pathways not normally active under physiological conditions.

2.7. Stromal-epithelial interactions

The signaling pathways described above focus on those events occurring within the secretory epithelium that are responsible for initiating cell death during post-lactational involution. However, it is becoming more apparent that stromal cells contribute substantially to post-lactational apoptosis [99-103]. This was recently demonstrated in a transgenic mouse model referred to as MAFIA (macrophage Fas-induced apoptosis) [104]. Macrophages from MAFIA mice express a modified Fas receptor that, in response to a dimerization-inducing small molecule (AP20187), triggers Fas-mediated apoptosis. Depletion of macrophages immediately prior to weaning impaired apoptosis within the secretory mammary epithelium, despite milk stasis and STAT3 activation [105]. These results demonstrate that macrophages are necessary to initiate apoptosis in the mammary epithelium. The underlying mechanism remains unclear at this point. However, it is possible that macrophages respond to signals emanating from mechanically stressed epithelia by producing factors that may activate the signaling pathways necessary for induction of apoptosis, or repress signaling pathways that may otherwise limit apoptosis. The transcriptional signatures generated from mouse mammary glands during involution were derived from whole tissue RNA, which would include not only epithelia but also the dynamic stromal components of the mammary gland. Therefore, it is possible that many expression events detected during early post-lactational involution are occurring within macrophage populations.

Mast cells are also heavily recruited to the mammary gland during post-lactational involution [106], and like macrophages, are critical for epithelial apoptosis during mammary gland involution [107]. Specifically, mast cells produce plasma kallikrien (PKal), the primary activator of plasminogen in the mammary gland. Expression of PKal rapidly increases during involution, and while PKal, plasminogen, and other serine proteases undoubtedly have a major role in tissue remodeling during later stages of involution, evidence suggests that PKal also drives epithelial apoptosis. Inhibition of mast cell-derived PKal during post-lactational involution impaired epithelial cell death, suggesting that mast cells are vital for triggering apoptosis in the post-lactational mammary gland. Interestingly, in the absence of STAT3 within the mammary epithelium, mast cells and macrophages are not recruited to the mammary gland during post-lactational involution [108], suggesting that recruitment of stromal cells to the involuting mammary gland is initiated by early apoptotic signaling events occurring within the epithelial compartment (**Fig. 1**).

2.8. Lysosomal membrane polarization

Although most studies suggest that mammary gland involution occurs by apoptosis, it has been proposed recently that several morphological features of the involuting mammary gland may resemble necrosis rather than apoptosis [109]. These include cytoplasmic swelling, lack of membrane blebbing, and lack of nuclear fragmentation. Using mice deficient for both caspase 3 and 6, it was shown that mammary gland involution could proceed in the absence of these two classical activators of apoptosis, suggesting that perhaps alternative mechanisms of programmed cell death may exist in the post-lactational mammary gland. The authors proposed that STAT3 activity could upregulate expression of lysosomal cathepsins, which may leak from lysosomes to activate cell death pathways [110, 111]. In support of this idea, cathepsin L is upregulated strongly with the onset of mammary involution [112]. Mice treated with a specific cathepsin L inhibitor during the first three days of involution demonstrated reduced cell death as compared to untreated mice [112]. Cathepsin-induced cell death can be simulated by ectopic addition of reactive oxygen species (ROS) to cultures of mammary epithelial cells. Interestingly, the ROS nitric oxide (NO) can trigger mammary gland involution after weaning [113] in mice. While the role of apoptotic cell death in the mammary gland is widely accepted, investigators should be aware of alternative cell death pathways that contribute to programmed cell death during involution.

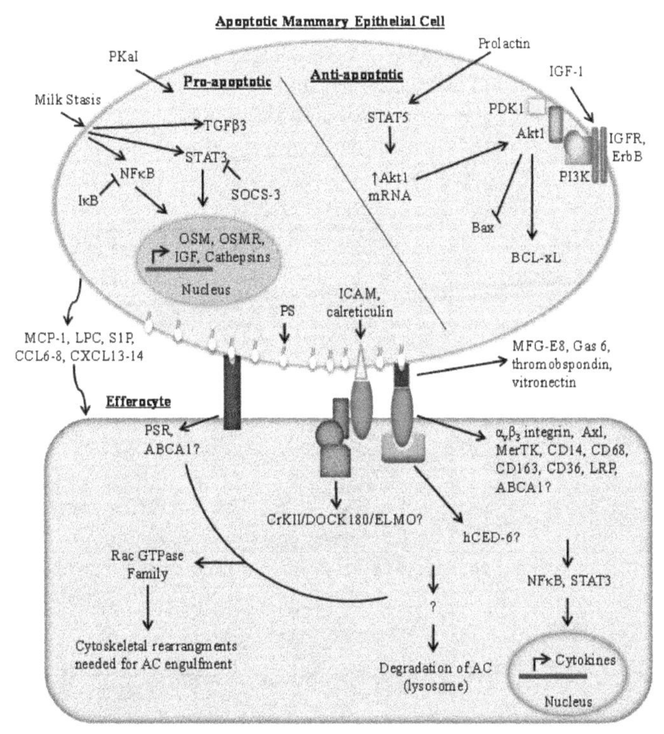

Figure 1. Apoptosis in the post-lactational mammary epithelial cell (MEC) is initiated by three molecular signals that are each required: TGFβ3, NFκB, and STAT3. Their loss impairs post-lactational apoptosis, despite continued milk stasis. Apoptotic MECs are cleared from the mammary gland by efferocytosis, or the phagocytic engulfment of dying cells. In the post-lactational mammary gland, MECs and macrophages engulf neighboring apoptotic cells. The phagocyte uses cell surface receptors to recognize, bind, and engulf apoptotic cells. These receptors include MerTK, Axl, αvβ3 integrin and others. Intracellular signaling pathways that regulate cytoskeletal rearrangements (such as Rac signaling) are necessary for efferocytosis. Once engulfed into vesicles, apoptotic cells are degraded by the lysosomal pathway. Efferocytosis activates NF-κB and STAT3, upregulating cytokines that are critical for post-lactational mammary remodeling.

2.9. Pathologies of the breast due to aberrant regulation of apoptosis

In the clinical setting, post-lactational involution of the secretory epithelium begins with milk stasis, at which point the secretory cells undergo apoptosis. Clearance of dying cells and residual milk is accomplished by phagocytes within the breast [114]. Regrowth of stromal adipose tissue and continued tissue remodeling returns the breast to a relatively quiescent state comprised of morphological structures similar to those found in nulliparous women. Rarely, the process of involution may be delayed. Failure to remove unnecessary lactational cells may result in symptomatic inflammatory tissue damage. Delayed involution in the human breast is characterized by the maintenance of secretory structures, loss of post-lactational apoptosis, and infiltration of the breast by inflammatory cells. Focal calcification may also be present [5]. Ductal distention of accumulated milk can be painful. Stagnant milk can be a source for infection and mastitis, to which the gland would respond with secretion of acute inflammatory cytokines and recruitment of leukocytes [115]. Similar to this clinical scenario, mouse models of delayed post-lactational cell death commonly develop mastitis [49, 108, 116].

EMS receptor	Bridging Molecule (BM)	EMS	Mammary Tissue and Cell Line Expression
PSR	-	PS [117]	Primary mouse mammary epithelia [117]
TIMs	-	PS [118-124]	unknown
BAI1	-	PS [125]	unknown
Stabilin-2	-	PS [126, 127].	unknown
ABCA1	unknown	PS [128, 129]	Bovine [130], mouse, human mammary [129]
$\alpha_v\beta_3$ integrin	Vitronectin [131] Thrombospondin [132] MFG-E8 [133]	PS [131, 132]	Human MCF10A, MCF-7, and MDA-MB-231 cells [134] bovine [135], mouse mammary gland [136, 137]
Tyro3	Gas6 [138] Protein S [139, 140]	PS [138]	unknown
Axl	Gas6 [138, 141]	PS [138, 141]	Human breast [142]
MerTK	Gas6 [138] Protein S[140]	PS [138, 140]	Mouse mammary [116]
CR3/CR4	unknown	C3bi [143]	unknown
CD14	unknown	ICAM [144]	Bovine [145], canine [146], mouse [4], and human mammary [144, 147]
CD68	unknown	unknown	Human, mouse mammary macrophages [148]
CD163	unknown	unknown	Human breast [148]
CD36	Thrombospondin [132]	unknown	MDA-MB-435, MDA-MB-231 human cells in mouse mammary [149]
LRP	β_2GP1 [150] C1q [151]	PS [150] Calreticulin [151]	Rat mammary gland [152] Normal, transformed mammary epithelia [153]
Marco [154]	unknown	unknown	unknown

Table 1. Key Factors Involved in Efferocytosis

Recent data garnered from mouse models of delayed post-lactational involution suggest that deregulation of post-lactational apoptosis may facilitate mammary tumor formation [34, 70, 73, 155-157]. Observations made in human populations also suggest that altered post-lactational involution may associate with tumor formation in the breast [100-102, 158]. This may reflect micro-environmental influences, or may be a function cell death-dependent removal of unnecessary breast epithelial cells in a regulated fashion. In many cancers, intrinsic cell deaths mechanisms become suppressed, contributing to the net growth of the transformed cell population. For example, activation of STAT5 in post-lactational mouse mammary glands delays apoptosis, and results in formation of mammary tumors that express estrogen and progesterone receptors (ER+PR+), as well as activated STAT3 and STAT5 [159]. Moreover, post-lactational transcriptional programs initiated by NF-κB and STAT3 not only support cell death, but also enhance tumor formation and progression by inducing expression of pro-tumorigenic cytokines [23, 24, 28, 34, 160-162]. Indeed, the transiently increased risk of developing breast cancer in the five years following a pregnancy may be greatly influenced by a deregulated tumor microenvironment developed in the post-lactational breast [101].

While post-lactational involution and age-related lobular involution are distinct processes, recent studies indicate that both are related to breast cancer development. Clinical studies show that completion of lobular involution may reduce future breast cancer incidence [163-166]. With aging, there is a gradual loss of breast epithelial tissue that typically begins in peri-menopause, which then accelerates during menopause. Lobular involution is characterized by the apoptosis-mediated decrease in the size and complexity of the ductal tree and of the TDLU. This is distinct from post-lactational involution, which occurs very rapidly by comparison. However, similar mechanisms controlling apoptosis of the breast epithelium may occur in these two distinct models of involution.

Lobular involution, like post-lactational involution, may inversely correlate with breast cancer risk, since premenopausal women who underwent partial or complete lobular involution had a substantially decreased incidence of breast cancer, while postmenopausal women who showed delayed lobular involution were found to have a correspondingly elevated breast cancer incidence [164]. While much remains to be learned about how lobular involution is regulated, some clinical studies and animal models suggest that IGF-1 may inhibit involution of lobules in the breast [163]. Clinically, a cross-sectional study among 472 women demonstrated that higher IGF-1 levels associated with incomplete lobular involution, supporting the idea that IGF-1/PI3K/Akt-induced survival pathways prevent physiologic cell death, leading to pathological consequences.

2.10. Introduction to efferocytosis in the breast

Following apoptosis, one final event is needed to truly complete the life of the cell. This final step is phagocytic engulfment of apoptotic cells or 'efferocytosis'. The term 'efferocytosis' was recently coined by Hensen et al. to distinguish phagocytic apoptotic cell removal from phagocytic pathogen removal [167]. While both processes are executed by phagocytes, they result in distinctly different biological responses, one characterized by a dampened acute inflammatory response and upregulation of tolerogenic and wound healing effectors (efferocytosis), while the other is characterized by a pro-inflammatory response (pathogen removal).

Efferocytosis is a carefully regulated process involving recruitment of phagocytes to the apoptotic cell, recognition of the apoptotic cell by the phagocytes, engulfment of the apoptotic cell by the phagocyte, and final breakdown of apoptotic cell components. If disrupted, apoptotic cells will undergo necrotic lysis, leading to acute inflammation, tissue damage and autoimmunity. Therefore, efferocytosis is critical for tissue homeostasis. However, recent discoveries indicate that the normal process of efferocytosis may be undesirable under certain pathological conditions, such as in the tumor microenvironment. We will discuss apoptotic cell clearance in the normal post-lactational breast and in the breast tumor microenvironment.

2.11. The process of efferocytosis

In general, clearance of apoptotic cells is often executed by macrophages and dendritic cells (DCs), but can also be performed by fibroblasts, endothelial and epithelial cells. A cell that engulfs an apoptotic cell through phagocytic mechanisms is called an efferocyte, regardless of its origin. Studies performed in cell culture and *in vivo* demonstrate that MECs and macrophages are both capable of engulfing apoptotic MECs during post-lactational involution of the secretory mammary epithelium [114, 116, 168].

Macrophages, the 'professional phagocytes' of the immune system, must infiltrate the mammary gland in response to the physiological presence of apoptotic cells during involution. Large quantities of bone marrow-derived and spleen-derived macrophages infiltrate the post-lactational mouse mammary gland in response to STAT3 activation and apoptosis of the mammary secretory epithelium. It is thought that apoptotic cells may release soluble chemo-attractants, or 'find me' signals, which recruit macrophages to the post-lactational mammary gland. For example, Monocyte Chemo-attractant Protein-1 (MCP-1/CCL2) is released from apoptotic cells in an NF-κB-dependent manner [169]. Interestingly, MCP-1 expression is strongly induced in the mammary gland at day 2 of involution, a time point that follows NF-κB-induced cell death, and that precedes macrophage influx in the post-lactational mammary gland. These observations are consistent with NF-κB-induced apoptosis followed by NF-κB-induced expression of an efferocyte chemo-attractant, although this has not yet been demonstrated. Additional chemokines including CX3CL1, CCL6, CCL7, CCL8, and CXCL14, are induced during post-lactational involution and may be signals that recruit macrophages to the involuting mammary gland to clear the accumulating apoptotic cell burden [170, 171] [4, 16].

Histological evidence of apoptotic cells within cytoplasmic vacuoles of mammary macrophages confirms that mammary macrophages engulf apoptotic cells during involution. Once present within the mammary gland, the macrophage identifies apoptotic cells, scanning for signals that are present on the apoptotic cell but not a healthy cell, often referred to as an 'eat me' signal (EMS, **Table 1**). The earliest and most recognized EMS is surface exposure of phosphatidylserine (PS). Healthy cells actively maintain PS on the inner plasma membrane leaflet. At the onset of apoptosis, PS is presented to the outer leaflet thus acting as a marker for a dying cell that requires engulfment [172-174]. EMSs are recognized by macrophages that express EMS receptors on their cell surface. EMS receptors may bind directly to the EMS on the dying cell. For example, the PS receptor (PSR), a transmembrane protein expressed by macrophages [128] and MECs [129], directly binds PS. Brain angiogenesis inhibitor 1 (BAI1) also binds PS directly, and is important for macrophage-mediated efferocytosis [125], but has not yet been studied in the mammary gland. Stabilin-2 and members of the T cell immunoglobin and mucin (TIM) family of receptors also directly bind PS [118-124, 126, 127], demonstrating mechanistic redundancy in the efferocytic pathways. It should be noted, however, that PS is an insufficient EMS, as macrophages fail to recognize live cells in which PS is forced to the outer leaflet [173, 175].

While some EMS receptors bind apoptotic cell EMSs directly, other EMS receptors bind an extracellular bridging molecule that simultaneously binds the EMS and the EMS receptor (**Table 1**). For example, the bridging ligand milkfat globule epidermal growth factor-like 8 (MFG-E8) binds to PS on apoptotic cells [133, 176, 177], while binding $\alpha_v\beta_3$ and $\alpha_v\beta_5$ integrins expressed by macrophages. Growth arrest specific gene 6 (Gas6) and Protein S [178, 179] are bridging molecules that bind to the EMS receptors MerTK, Axl, and Tyro3 [180] expressed by macrophages, while simultaneously binding PS on the apoptotic cell.

Once the apoptotic cell is bound to the macrophage, intracellular signaling pathways must remodel the actin cytoskeleton to drive phagocytic ingestion of the apoptotic cell. Most of these events have been mapped out in *Caenorhabditis elegans* (*C. elegans*), which are discussed in detail within comprehensive reviews by Reddien et al. and Ravichandran et al. [181, 182]. In mammalian macrophages, intracellular signaling networks that regulate actin cytoskeletal dynamics are required for apoptotic cell engulfment. For example, a protein complex comprised of CrkII, DOCK180, and ELMO causes activation of the Rac GTPase, a master regulator of actin cytoskeletal dynamics. Actin-dependent membrane extensions physically engulf the apoptotic cell [183-187]. Rac-mediated actin rearrangements are countered by RhoA [188], which prevents the formation of cell extensions needed for efferocytosis [189]. The engulfed apoptotic cell is then consumed by the efferocyte, primarily through lysosomal degradation. Many signaling factors that drive apoptotic cell engulfment also enhance lysosomal degradation of apoptotic cells. For example, Rac is essential for maturation of phago-lysosomes in macrophages [190-193].

In addition to efferocytosis in the mammary gland, macrophages are also critical for cytokine modulation and extracellular matrix remodeling during the second phase of involution, underscoring the important role of macrophages in the post-lactational breast. Given their known role as professional phagocytes and their massive influx to the post-lactational mammary gland, it is perhaps not surprising that efferocytosis by macrophages occurs in the post-lactational breast. What is more surprising is that apoptotic cell clearance occurs on a profound scale prior to the influx of macrophages to the involuting mammary gland. Recent evidence demonstrated that MECs are the primary efferocytes of the breast during the earliest stages of post-lactational involution, the first three days prior to the influx of macrophages. The ability of MECs to act as efferocytes ensures a rapid response to the massive level of apoptosis that occurs during post-lactational involution. Loss of MEC-mediated efferocytosis impairs post-lactational homeostasis, resulting in chronic mammary inflammation, scarring and inhibition of future lactation.

Interestingly, MECs utilize many of the same EMS receptors used by macrophages to recognize apoptotic MECs. For example, MerTK is critical for MEC-mediated efferocytosis during post-lactational involution. MerTK loss from the mouse mammary epithelium causes apoptotic cell accumulation and milk stasis [116], despite the presence of wild-type macrophages. Interestingly, mRNA and protein expression of MerTK is dramatically upregulated by post-lactational day 1 within the luminal mammary epithelium. Similarly, the integrins $\alpha_v\beta_3$ and $\alpha_v\beta_5$ are expressed in the early post-lactational mammary epithelium. The $\alpha_v\beta_5$ bridging ligand, MFG-E8, is simultaneously induced [194], increasing the physical interaction between MFG-E8 and PS [136] and driving the clearance of membrane-coated milk components from the involuting mammary gland [195].

2.12. Physiological and pathological consequences of efferocytosis

After the efferocyte removes the apoptotic cell, transcriptional events in the efferocyte result in cytokine, chemokine and growth factor production. The combined profile of the factors produced by the efferocyte promotes wound healing through enhanced tissue remodeling, angiogenesis, proliferation and resolution of acute inflammation. The efferocytosis-induced wound healing cytokine profile contrasts sharply to the cytokine profile produced in response to phagocytosis of pathogens, which is characterized by acute inflammatory cytokines [196]. In fact, efferocytosis is thought to be a key step in resolving or dampening acute inflammatory cytokine expression following tissue injury or pathogen exposure, resulting in repair and homeostasis [197]. Microarray analyses of mammary glands harvested at early post-lactational involution time points displayed a pronounced wound healing expression signature [3, 4, 158, 168, 198, 199], consistent with transcriptional changes that result from efferocytosis. The prominent role of efferocytosis in re-establishing mammary homeostasis following widespread apoptosis of the secretory epithelium was shown by experiments in which loss of efferocytosis resulted in apoptotic cell accumulation, sustained milk stasis within ductal lumens, inflammation and scarring [116]. These pathological changes impaired lactation in future pregnancies.

Although key to re-establishing homeostasis in the post-lactational mammary gland [116], recent evidence indicates that efferocytosis may support a more malignant tumor microenvironment [101, 102, 105, 106, 158, 198, 200-202]. Researchers are beginning to address the overlapping roles of efferocytic and metastasis-promoting cytokines. In agreement with this idea, mouse studies show that breast tumors grew more rapidly, invaded more readily, and formed distant metastases more efficiently when implanted in the post-lactational mammary gland as compared to implantation into a nulliparous mammary gland [101]. One explanation for this observation is that post-lactational efferocytosis promotes breast tumor malignancy through production of wound healing cytokines, which are known to drive breast cancer growth and invasion. In support of this idea, MFG-E8 [203] and its receptor $\alpha_v\beta_{3/5}$, as well as Gas6 and its ligand MerTK [116, 204] are frequently overexpressed in breast cancers. One recent study demonstrated that the pro-tumorigenic cytokine IL-6 induces expression of MerTK, enhancing the ability of macrophages to engulf apoptotic cells and increasing production of wound healing cytokines such as IL-4 and IL-10 [205]. Recently published data implicates MerTK in breast cancer metastasis [206].

This observation has clinical relevance to pregnancy associated breast cancers (PABCs), defined as breast cancers that arise during the 5 years following a pregnancy. PABCs are among the most malignantly aggressive breast cancers, and are thus associated with poor prognosis. A better understanding of the processes outline above will undoubtedly expand the therapeutic options for these patients.

3. Conclusion

Altogether, these data support the hypothesis that targeting mediators of efferocytosis may limit pro-tumorigenic cytokine production. Moreover, it is becoming increasingly apparent that many factors within the mammary gland cooperate to ensure apoptosis and apoptotic cell clearance, highlighting the complexity of these processes and the need for more detailed investigations. Due to the dominant role of apoptosis and efferocytosis in maintaining tissue homeostasis, especially during post-lactational involution, the mammary gland provides an ideal platform for future study.

Author details

Jamie C. Stanford and Rebecca S. Cook
Department of Cancer Biology, Vanderbilt University, Nashville, TN, USA

Acknowledgement

This work was supported by NIH R01 CA143126 (RSC), and Susan G. Komen KG100677 (RSC).

4. References

[1] Humphreys, R.C., et al., Apoptosis in the terminal endbud of the murine mammary gland: a mechanism of ductal morphogenesis. Development, 1996. 122(12): p. 4013-22.

[2] Andres, A.C. and R. Strange, Apoptosis in the estrous and menstrual cycles. J Mammary Gland Biol Neoplasia, 1999. 4(2): p. 221-8.

[3] Atabai, K., D. Sheppard, and Z. Werb, Roles of the innate immune system in mammary gland remodeling during involution. J Mammary Gland Biol Neoplasia, 2007. 12(1): p. 37-45.

[4] Stein, T., et al., Involution of the mouse mammary gland is associated with an immune cascade and an acute-phase response, involving LBP, CD14 and STAT3. Breast Cancer Res, 2004. 6(2): p. R75-91.

[5] de Almeida, C.J. and R. Linden, Phagocytosis of apoptotic cells: a matter of balance. Cell Mol Life Sci, 2005. 62(14): p. 1532-46.

[6] Hinck, L. and G.B. Silberstein, Key stages in mammary gland development: the mammary end bud as a motile organ. Breast Cancer Res, 2005. 7(6): p. 245-51.

[7] Mailleux, A.A., et al., BIM regulates apoptosis during mammary ductal morphogenesis, and its absence reveals alternative cell death mechanisms. Dev Cell, 2007. 12(2): p. 221-34.

[8] Li, M., et al., Mammary-derived signals activate programmed cell death during the first stage of mammary gland involution. Proc Natl Acad Sci U S A, 1997. 94(7): p. 3425-30.

[9] Marti, A., et al., Milk accumulation triggers apoptosis of mammary epithelial cells. Eur J Cell Biol, 1997. 73(2): p. 158-65.

[10] Quaglino, A., et al., Mechanical strain induces involution-associated events in mammary epithelial cells. BMC Cell Biol, 2009. 10: p. 55.

[11] VanHouten, J.N., Calcium sensing by the mammary gland. J Mammary Gland Biol Neoplasia, 2005. 10(2): p. 129-39.

[12] VanHouten, J.N., M.C. Neville, and J.J. Wysolmerski, The calcium-sensing receptor regulates plasma membrane calcium adenosine triphosphatase isoform 2 activity in mammary epithelial cells: a mechanism for calcium-regulated calcium transport into milk. Endocrinology, 2007. 148(12): p. 5943-54.

[13] VanHouten, J., et al., PMCA2 regulates apoptosis during mammary gland involution and predicts outcome in breast cancer. Proc Natl Acad Sci U S A, 2010. 107(25): p. 11405-10.

[14] Tremblay, G., et al., Local control of mammary involution: is stanniocalcin-1 involved? J Dairy Sci, 2009. 92(5): p. 1998-2006.

[15] Hasilo, C.P., et al., Nuclear targeting of stanniocalcin to mammary gland alveolar cells during pregnancy and lactation. Am J Physiol Endocrinol Metab, 2005. 289(4): p. E634-42.

[16] Clarkson, R.W., et al., Gene expression profiling of mammary gland development reveals putative roles for death receptors and immune mediators in post-lactational regression. Breast Cancer Res, 2004. 6(2): p. R92-109.

[17] Chapman, R.S., et al., The role of Stat3 in apoptosis and mammary gland involution. Conditional deletion of Stat3. Adv Exp Med Biol, 2000. 480: p. 129-38.

[18] Chapman, R.S., et al., Suppression of epithelial apoptosis and delayed mammary gland involution in mice with a conditional knockout of Stat3. Genes Dev, 1999. 13(19): p. 2604-16.

[19] Humphreys, R.C., et al., Deletion of Stat3 blocks mammary gland involution and extends functional competence of the secretory epithelium in the absence of lactogenic stimuli. Endocrinology, 2002. 143(9): p. 3641-50.

[20] Song, L., et al., Activation of Stat3 by receptor tyrosine kinases and cytokines regulates survival in human non-small cell carcinoma cells. Oncogene, 2003. 22(27): p. 4150-65.

[21] Zhong, Z., Z. Wen, and J.E. Darnell, Jr., Stat3: a STAT family member activated by tyrosine phosphorylation in response to epidermal growth factor and interleukin-6. Science, 1994. 264(5155): p. 95-8.

[22] Pensa, S., C.J. Watson, and V. Poli, Stat3 and the inflammation/acute phase response in involution and breast cancer. J Mammary Gland Biol Neoplasia, 2009. 14(2): p. 121-9.

[23] Dauer, D.J., et al., Stat3 regulates genes common to both wound healing and cancer. Oncogene, 2005. 24(21): p. 3397-408.

[24] Aggarwal, B.B., et al., Signal transducer and activator of transcription-3, inflammation, and cancer: how intimate is the relationship? Ann N Y Acad Sci, 2009. 1171: p. 59-76.

[25] Aoki, Y., G.M. Feldman, and G. Tosato, Inhibition of STAT3 signaling induces apoptosis and decreases survivin expression in primary effusion lymphoma. Blood, 2003. 101(4): p. 1535-42.

[26] Kanda, N., et al., STAT3 is constitutively activated and supports cell survival in association with survivin expression in gastric cancer cells. Oncogene, 2004. 23(28): p. 4921-9.

[27] Catlett-Falcone, R., et al., Constitutive activation of Stat3 signaling confers resistance to apoptosis in human U266 myeloma cells. Immunity, 1999. 10(1): p. 105-15.

[28] Bromberg, J.F., et al., Stat3 as an oncogene. Cell, 1999. 98(3): p. 295-303.

[29] Sutherland, K.D., G.J. Lindeman, and J.E. Visvader, Knocking off SOCS genes in the mammary gland. Cell Cycle, 2007. 6(7): p. 799-803.

[30] Sutherland, K.D., et al., c-myc as a mediator of accelerated apoptosis and involution in mammary glands lacking Socs3. EMBO J, 2006. 25(24): p. 5805-15.

[31] Thangaraju, M., et al., C/EBPdelta is a crucial regulator of pro-apoptotic gene expression during mammary gland involution. Development, 2005. 132(21): p. 4675-85.

[32] Thangaraju, M., S. Sharan, and E. Sterneck, Comparison of mammary gland involution between 129S1 and C57BL/6 inbred mouse strains: differential regulation of Bcl2a1, Trp53, Cebpb, and Cebpd expression. Oncogene, 2004. 23(14): p. 2548-53.

[33] Tiffen, P.G., et al., A dual role for oncostatin M signaling in the differentiation and death of mammary epithelial cells in vivo. Mol Endocrinol, 2008. 22(12): p. 2677-88.

[34] Abdulghani, J., et al., Stat3 promotes metastatic progression of prostate cancer. Am J Pathol, 2008. 172(6): p. 1717-28.

[35] Ranger, J.J., et al., Identification of a Stat3-dependent transcription regulatory network involved in metastatic progression. Cancer Res, 2009. 69(17): p. 6823-30.

[36] Kataoka, K., et al., Stage-specific disruption of Stat3 demonstrates a direct requirement during both the initiation and promotion stages of mouse skin tumorigenesis. Carcinogenesis, 2008. 29(6): p. 1108-14.

[37] Yeh, H.H., et al., Ha-ras oncogene-induced Stat3 phosphorylation enhances oncogenicity of the cell. DNA Cell Biol, 2009. 28(3): p. 131-9.

[38] Yeh, H.H., et al., Autocrine IL-6-induced Stat3 activation contributes to the pathogenesis of lung adenocarcinoma and malignant pleural effusion. Oncogene, 2006. 25(31): p. 4300-9.

[39] Baxter, F.O., et al., IKKbeta/2 induces TWEAK and apoptosis in mammary epithelial cells. Development, 2006. 133(17): p. 3485-94.

[40] Karin, M. and Y. Ben-Neriah, Phosphorylation meets ubiquitination: the control of NF-[kappa]B activity. Annu Rev Immunol, 2000. 18: p. 621-63.

[41] Brantley, D.M., et al., Dynamic expression and activity of NF-kappaB during post-natal mammary gland morphogenesis. Mech Dev, 2000. 97(1-2): p. 149-55.

[42] Li, Z.W., et al., The IKKbeta subunit of IkappaB kinase (IKK) is essential for nuclear factor kappaB activation and prevention of apoptosis. J Exp Med, 1999. 189(11): p. 1839-45.

[43] Maeda, S., et al., IKKbeta is required for prevention of apoptosis mediated by cell-bound but not by circulating TNFalpha. Immunity, 2003. 19(5): p. 725-37.

[44] Senftleben, U., et al., IKKbeta is essential for protecting T cells from TNFalpha-induced apoptosis. Immunity, 2001. 14(3): p. 217-30.

[45] Romashkova, J.A. and S.S. Makarov, NF-kappaB is a target of AKT in anti-apoptotic PDGF signalling. Nature, 1999. 401(6748): p. 86-90.

[46] Fujioka, S., et al., Stabilization of p53 is a novel mechanism for proapoptotic function of NF-kappaB. J Biol Chem, 2004. 279(26): p. 27549-59.

[47] Ashkenazi, A. and V.M. Dixit, Death receptors: signaling and modulation. Science, 1998. 281(5381): p. 1305-8.

[48] Baetu, T.M., et al., Disruption of NF-kappaB signaling reveals a novel role for NF-kappaB in the regulation of TNF-related apoptosis-inducing ligand expression. J Immunol, 2001. 167(6): p. 3164-73.

[49] Connelly, L., et al., Activation of nuclear factor kappa B in mammary epithelium promotes milk loss during mammary development and infection. J Cell Physiol, 2010. 222(1): p. 73-81.

[50] Engelman, J.A., J. Luo, and L.C. Cantley, The evolution of phosphatidylinositol 3-kinases as regulators of growth and metabolism. Nat Rev Genet, 2006. 7(8): p. 606-19.

[51] Wong, K.K., J.A. Engelman, and L.C. Cantley, Targeting the PI3K signaling pathway in cancer. Curr Opin Genet Dev, 2010. 20(1): p. 87-90.

[52] Wickenden, J.A. and C.J. Watson, Key signalling nodes in mammary gland development and cancer. Signalling downstream of PI3 kinase in mammary epithelium: a play in 3 Akts. Breast Cancer Res, 2010. 12(2): p. 202.

[53] Cantley, L.C., The phosphoinositide 3-kinase pathway. Science, 2002. 296(5573): p. 1655-7.

[54] Manning, B.D. and L.C. Cantley, AKT/PKB signaling: navigating downstream. Cell, 2007. 129(7): p. 1261-74.

[55] Vivanco, I. and C.L. Sawyers, The phosphatidylinositol 3-Kinase AKT pathway in human cancer. Nat Rev Cancer, 2002. 2(7): p. 489-501.
[56] Chipuk, J.E., et al., Mechanism of apoptosis induction by inhibition of the anti-apoptotic BCL-2 proteins. Proc Natl Acad Sci U S A, 2008. 105(51): p. 20327-32.
[57] Kuwana, T. and D.D. Newmeyer, Bcl-2-family proteins and the role of mitochondria in apoptosis. Curr Opin Cell Biol, 2003. 15(6): p. 691-9.
[58] Datta, S.R., et al., Akt phosphorylation of BAD couples survival signals to the cell-intrinsic death machinery. Cell, 1997. 91(2): p. 231-41.
[59] Heermeier, K., et al., Bax and Bcl-xs are induced at the onset of apoptosis in involuting mammary epithelial cells. Mech Dev, 1996. 56(1-2): p. 197-207.
[60] Schorr, K., et al., Bcl-2 gene family and related proteins in mammary gland involution and breast cancer. J Mammary Gland Biol Neoplasia, 1999. 4(2): p. 153-64.
[61] Metcalfe, A.D., et al., Developmental regulation of Bcl-2 family protein expression in the involuting mammary gland. J Cell Sci, 1999. 112 (Pt 11): p. 1771-83.
[62] Zeng, X., H. Xu, and R.I. Glazer, Transformation of mammary epithelial cells by 3-phosphoinositide-dependent protein kinase-1 (PDK1) is associated with the induction of protein kinase Calpha. Cancer Res, 2002. 62(12): p. 3538-43.
[63] Pinner, S. and E. Sahai, PDK1 regulates cancer cell motility by antagonising inhibition of ROCK1 by RhoE. Nat Cell Biol, 2008. 10(2): p. 127-37.
[64] Alessi, D.R., et al., Characterization of a 3-phosphoinositide-dependent protein kinase which phosphorylates and activates protein kinase Balpha. Curr Biol, 1997. 7(4): p. 261-9.
[65] Rucker, E.B., 3rd, et al., Forced involution of the functionally differentiated mammary gland by overexpression of the pro-apoptotic protein bax. Genesis, 2011. 49(1): p. 24-35.
[66] Walton, K.D., et al., Conditional deletion of the bcl-x gene from mouse mammary epithelium results in accelerated apoptosis during involution but does not compromise cell function during lactation. Mech Dev, 2001. 109(2): p. 281-93.
[67] Schorr, K., et al., Gain of Bcl-2 is more potent than bax loss in regulating mammary epithelial cell survival in vivo. Cancer Res, 1999. 59(11): p. 2541-5.
[68] Renner, O., et al., Activation of phosphatidylinositol 3-kinase by membrane localization of p110alpha predisposes mammary glands to neoplastic transformation. Cancer Res, 2008. 68(23): p. 9643-53.
[69] Dillon, R.L., et al., Akt1 and akt2 play distinct roles in the initiation and metastatic phases of mammary tumor progression. Cancer Res, 2009. 69(12): p. 5057-64.
[70] Schwertfeger, K.L., M.M. Richert, and S.M. Anderson, Mammary gland involution is delayed by activated Akt in transgenic mice. Mol Endocrinol, 2001. 15(6): p. 867-81.
[71] Maroulakou, I.G., et al., Distinct roles of the three Akt isoforms in lactogenic differentiation and involution. J Cell Physiol, 2008. 217(2): p. 468-77.
[72] Creamer, B.A., et al., Stat5 promotes survival of mammary epithelial cells through transcriptional activation of a distinct promoter in Akt1. Mol Cell Biol, 2010. 30(12): p. 2957-70.

[73] Neilson, L.M., et al., Coactivation of janus tyrosine kinase (Jak)1 positively modulates prolactin-Jak2 signaling in breast cancer: recruitment of ERK and signal transducer and activator of transcription (Stat)3 and enhancement of Akt and Stat5a/b pathways. Mol Endocrinol, 2007. 21(9): p. 2218-32.

[74] Flint, D.J., et al., Prolactin inhibits cell loss and decreases matrix metalloproteinase expression in the involuting mouse mammary gland but fails to prevent cell loss in the mammary glands of mice expressing IGFBP-5 as a mammary transgene. J Mol Endocrinol, 2006. 36(3): p. 435-48.

[75] Furth, P.A., et al., Signal transducer and activator of transcription 5 as a key signaling pathway in normal mammary gland developmental biology and breast cancer. Breast Cancer Res, 2011. 13(5): p. 220.

[76] Heidegger, I., et al., Targeting the insulin-like growth factor network in cancer therapy. Cancer Biol Ther, 2011. 11(8): p. 701-7.

[77] Flint, D.J., et al., Role of insulin-like growth factor binding proteins in mammary gland development. J Mammary Gland Biol Neoplasia, 2008. 13(4): p. 443-53.

[78] Sureshbabu, A., E. Tonner, and D.J. Flint, Insulin-like growth factor binding proteins and mammary gland development. Int J Dev Biol, 2011. 55(7-9): p. 781-9.

[79] Flint, D.J., et al., Insulin-like growth factor binding proteins initiate cell death and extracellular matrix remodeling in the mammary gland. Domest Anim Endocrinol, 2005. 29(2): p. 274-82.

[80] Lochrie, J.D., et al., Insulin-like growth factor binding protein (IGFBP)-5 is upregulated during both differentiation and apoptosis in primary cultures of mouse mammary epithelial cells. J Cell Physiol, 2006. 207(2): p. 471-9.

[81] Marshman, E., et al., Insulin-like growth factor binding protein 5 and apoptosis in mammary epithelial cells. J Cell Sci, 2003. 116(Pt 4): p. 675-82.

[82] Allar, M.A. and T.L. Wood, Expression of the insulin-like growth factor binding proteins during postnatal development of the murine mammary gland. Endocrinology, 2004. 145(5): p. 2467-77.

[83] Tonner, E., et al., Insulin-like growth factor binding protein-5 (IGFBP-5) potentially regulates programmed cell death and plasminogen activation in the mammary gland. Adv Exp Med Biol, 2000. 480: p. 45-53.

[84] Tonner, E., et al., Insulin-like growth factor binding protein-5 (IGFBP-5) induces premature cell death in the mammary glands of transgenic mice. Development, 2002. 129(19): p. 4547-57.

[85] Quaglino, A., et al., Mouse mammary tumors display Stat3 activation dependent on leukemia inhibitory factor signaling. Breast Cancer Res, 2007. 9(5): p. R69.

[86] Schere-Levy, C., et al., Leukemia inhibitory factor induces apoptosis of the mammary epithelial cells and participates in mouse mammary gland involution. Exp Cell Res, 2003. 282(1): p. 35-47.

[87] Kritikou, E.A., et al., A dual, non-redundant, role for LIF as a regulator of development and STAT3-mediated cell death in mammary gland. Development, 2003. 130(15): p. 3459-68.

[88] Matsuda, M., et al., Serotonin regulates mammary gland development via an autocrine-paracrine loop. Dev Cell, 2004. 6(2): p. 193-203.

[89] Song, J., et al., Roles of Fas and Fas ligand during mammary gland remodeling. J Clin Invest, 2000. 106(10): p. 1209-20.

[90] Sohn, B.H., et al., Interleukin-10 up-regulates tumour-necrosis-factor-alpha-related apoptosis-inducing ligand (TRAIL) gene expression in mammary epithelial cells at the involution stage. Biochem J, 2001. 360(Pt 1): p. 31-8.

[91] Faure, E., et al., Differential expression of TGF-beta isoforms during postlactational mammary gland involution. Cell Tissue Res, 2000. 300(1): p. 89-95.

[92] Robinson, S.D., et al., Regulated expression and growth inhibitory effects of transforming growth factor-beta isoforms in mouse mammary gland development. Development, 1991. 113(3): p. 867-78.

[93] Nguyen, A.V. and J.W. Pollard, Transforming growth factor beta3 induces cell death during the first stage of mammary gland involution. Development, 2000. 127(14): p. 3107-18.

[94] Yang, Y.A., et al., Smad3 in the mammary epithelium has a nonredundant role in the induction of apoptosis, but not in the regulation of proliferation or differentiation by transforming growth factor-beta. Cell Growth Differ, 2002. 13(3): p. 123-30.

[95] Gorska, A.E., et al., Transgenic mice expressing a dominant-negative mutant type II transforming growth factor-beta receptor exhibit impaired mammary development and enhanced mammary tumor formation. Am J Pathol, 2003. 163(4): p. 1539-49.

[96] Joseph, H., et al., Overexpression of a kinase-deficient transforming growth factor-beta type II receptor in mouse mammary stroma results in increased epithelial branching. Mol Biol Cell, 1999. 10(4): p. 1221-34.

[97] Bierie, B., et al., TGF-beta promotes cell death and suppresses lactation during the second stage of mammary involution. J Cell Physiol, 2009. 219(1): p. 57-68.

[98] Muraoka-Cook, R.S., et al., Activated type I TGFbeta receptor kinase enhances the survival of mammary epithelial cells and accelerates tumor progression. Oncogene, 2006. 25(24): p. 3408-23.

[99] Cunha, G.R. and Y.K. Hom, Role of mesenchymal-epithelial interactions in mammary gland development. J Mammary Gland Biol Neoplasia, 1996. 1(1): p. 21-35.

[100] Howlett, A.R. and M.J. Bissell, The influence of tissue microenvironment (stroma and extracellular matrix) on the development and function of mammary epithelium. Epithelial Cell Biol, 1993. 2(2): p. 79-89.

[101] Lyons, T.R., et al., Postpartum mammary gland involution drives progression of ductal carcinoma in situ through collagen and COX-2. Nat Med, 2011. 17(9): p. 1109-15.

[102] O'Brien, J., et al., Alternatively activated macrophages and collagen remodeling characterize the postpartum involuting mammary gland across species. Am J Pathol, 2010. 176(3): p. 1241-55.

[103] Reed, J.R. and K.L. Schwertfeger, Immune cell location and function during post-natal mammary gland development. J Mammary Gland Biol Neoplasia, 2010. 15(3): p. 329-39.

[104] Burnett, S.H., et al., Conditional macrophage ablation in transgenic mice expressing a Fas-based suicide gene. J Leukoc Biol, 2004. 75(4): p. 612-23.

[105] O'Brien, J., et al., Macrophages are crucial for epithelial cell death and adipocyte repopulation during mammary gland involution. Development, 2012. 139(2): p. 269-75.

[106] Ramirez, R.A., et al., Alterations in mast cell frequency and relationship to angiogenesis in the rat mammary gland during windows of physiologic tissue remodeling. Dev Dyn, 2012. 241(5): p. 890-900.

[107] Lilla, J.N., et al., Active plasma kallikrein localizes to mast cells and regulates epithelial cell apoptosis, adipocyte differentiation, and stromal remodeling during mammary gland involution. J Biol Chem, 2009. 284(20): p. 13792-803.

[108] Hughes, K., et al., Conditional deletion of Stat3 in mammary epithelium impairs the acute phase response and modulates immune cell numbers during post-lactational regression. J Pathol, 2012. 227(1): p. 106-17.

[109] Kreuzaler, P.A., et al., Stat3 controls lysosomal-mediated cell death in vivo. Nat Cell Biol, 2011. 13(3): p. 303-9.

[110] Margaryan, N.V., et al., New insights into cathepsin D in mammary tissue development and remodeling. Cancer Biol Ther, 2010. 10(5): p. 457-66.

[111] Lockshin, R.A. and Z. Zakeri, Caspase-independent cell death? Oncogene, 2004. 23(16): p. 2766-73.

[112] Burke, M.A., et al., Cathepsin L plays an active role in involution of the mouse mammary gland. Dev Dyn, 2003. 227(3): p. 315-22.

[113] Zaragoza, R., et al., Nitric oxide triggers mammary gland involution after weaning: remodelling is delayed but not impaired in mice lacking inducible nitric oxide synthase. Biochem J, 2010. 428(3): p. 451-62.

[114] Monks, J., et al., Epithelial cells remove apoptotic epithelial cells during post-lactation involution of the mouse mammary gland. Biol Reprod, 2008. 78(4): p. 586-94.

[115] Kumar, Y., et al., Delayed involution of lactation presenting as a non-resolving breast mass: a case report. J Med Case Reports, 2008. 2: p. 327.

[116] Sandahl, M., et al., Epithelial cell-directed efferocytosis in the post-partum mammary gland is necessary for tissue homeostasis and future lactation. BMC Dev Biol, 2010. 10: p. 122.

[117] Fadok, V.A., et al., A receptor for phosphatidylserine-specific clearance of apoptotic cells. Nature, 2000. 405(6782): p. 85-90.

[118] DeKruyff, R.H., et al., T cell/transmembrane, Ig, and mucin-3 allelic variants differentially recognize phosphatidylserine and mediate phagocytosis of apoptotic cells. J Immunol, 2010. 184(4): p. 1918-30.

[119] Ichimura, T., et al., Kidney injury molecule-1 is a phosphatidylserine receptor that confers a phagocytic phenotype on epithelial cells. J Clin Invest, 2008. 118(5): p. 1657-68.

[120] Kobayashi, N., et al., TIM-1 and TIM-4 glycoproteins bind phosphatidylserine and mediate uptake of apoptotic cells. Immunity, 2007. 27(6): p. 927-40.

[121] Miyanishi, M., et al., Identification of Tim4 as a phosphatidylserine receptor. Nature, 2007. 450(7168): p. 435-9.

[122] Rodriguez-Manzanet, R., et al., T and B cell hyperactivity and autoimmunity associated with niche-specific defects in apoptotic body clearance in TIM-4-deficient mice. Proc Natl Acad Sci U S A, 2010. 107(19): p. 8706-11.

[123] Santiago, C., et al., Structures of T Cell immunoglobulin mucin receptors 1 and 2 reveal mechanisms for regulation of immune responses by the TIM receptor family. Immunity, 2007. 26(3): p. 299-310.

[124] Wong, K., et al., Phosphatidylserine receptor Tim-4 is essential for the maintenance of the homeostatic state of resident peritoneal macrophages. Proc Natl Acad Sci U S A, 2010. 107(19): p. 8712-7.

[125] Park, D., et al., BAI1 is an engulfment receptor for apoptotic cells upstream of the ELMO/Dock180/Rac module. Nature, 2007. 450(7168): p. 430-4.

[126] Park, S.Y., et al., Epidermal growth factor-like domain repeat of stabilin-2 recognizes phosphatidylserine during cell corpse clearance. Mol Cell Biol, 2008. 28(17): p. 5288-98.

[127] Park, S.Y., et al., Rapid cell corpse clearance by stabilin-2, a membrane phosphatidylserine receptor. Cell Death Differ, 2008. 15(1): p. 192-201.

[128] Luciani, M.F. and G. Chimini, The ATP binding cassette transporter ABC1, is required for the engulfment of corpses generated by apoptotic cell death. EMBO J, 1996. 15(2): p. 226-35.

[129] Mani, O., et al., Expression, localization, and functional model of cholesterol transporters in lactating and nonlactating mammary tissues of murine, bovine, and human origin. Am J Physiol Regul Integr Comp Physiol, 2010. 299(2): p. R642-54.

[130] Mani, O., et al., Identification of ABCA1 and ABCG1 in milk fat globules and mammary cells--implications for milk cholesterol secretion. J Dairy Sci, 2011. 94(3): p. 1265-76.

[131] Yebra, M., et al., Requirement of receptor-bound urokinase-type plasminogen activator for integrin alphavbeta5-directed cell migration. J Biol Chem, 1996. 271(46): p. 29393-9.

[132] Savill, J., et al., Thrombospondin cooperates with CD36 and the vitronectin receptor in macrophage recognition of neutrophils undergoing apoptosis. J Clin Invest, 1992. 90(4): p. 1513-22.

[133] Akakura, S., et al., The opsonin MFG-E8 is a ligand for the alphavbeta5 integrin and triggers DOCK180-dependent Rac1 activation for the phagocytosis of apoptotic cells. Exp Cell Res, 2004. 292(2): p. 403-16.

[134] D'Mello, V., et al., The urokinase plasminogen activator receptor promotes efferocytosis of apoptotic cells. J Biol Chem, 2009. 284(25): p. 17030-8.

[135] Hvarregaard, J., et al., Characterization of glycoprotein PAS-6/7 from membranes of bovine milk fat globules. Eur J Biochem, 1996. 240(3): p. 628-36.

[136] Nakatani, H., et al., Weaning-induced expression of a milk-fat globule protein, MFG-E8, in mouse mammary glands, as demonstrated by the analyses of its mRNA, protein and phosphatidylserine-binding activity. Biochem J, 2006. 395(1): p. 21-30.

[137] Oshima, K., et al., Lactation-dependent expression of an mRNA splice variant with an exon for a multiply O-glycosylated domain of mouse milk fat globule glycoprotein MFG-E8. Biochem Biophys Res Commun, 1999. 254(3): p. 522-8.

[138] Nagata, K., et al., Identification of the product of growth arrest-specific gene 6 as a common ligand for Axl, Sky, and Mer receptor tyrosine kinases. J Biol Chem, 1996. 271(47): p. 30022-7.

[139] Stitt, T.N., et al., The anticoagulation factor protein S and its relative, Gas6, are ligands for the Tyro 3/Axl family of receptor tyrosine kinases. Cell, 1995. 80(4): p. 661-70.

[140] Prasad, D., et al., TAM receptor function in the retinal pigment epithelium. Mol Cell Neurosci, 2006. 33(1): p. 96-108.

[141] Varnum, B.C., et al., Axl receptor tyrosine kinase stimulated by the vitamin K-dependent protein encoded by growth-arrest-specific gene 6. Nature, 1995. 373(6515): p. 623-6.

[142] Berclaz, G., et al., Estrogen dependent expression of the receptor tyrosine kinase axl in normal and malignant human breast. Ann Oncol, 2001. 12(6): p. 819-24.

[143] Mevorach, D., et al., Complement-dependent clearance of apoptotic cells by human macrophages. J Exp Med, 1998. 188(12): p. 2313-20.

[144] Moffatt, O.D., et al., Macrophage recognition of ICAM-3 on apoptotic leukocytes. J Immunol, 1999. 162(11): p. 6800-10.

[145] Baravalle, C., et al., Proinflammatory cytokines and CD14 expression in mammary tissue of cows following intramammary inoculation of Panax ginseng at drying off. Vet Immunol Immunopathol, 2011. 144(1-2): p. 52-60.

[146] Krol, M., et al., Density of tumor-associated macrophages (TAMs) and expression of their growth factor receptor MCSF-R and CD14 in canine mammary adenocarcinomas of various grade of malignancy and metastasis. Pol J Vet Sci, 2011. 14(1): p. 3-10.

[147] Feng, A.L., et al., CD16+ monocytes in breast cancer patients: expanded by monocyte chemoattractant protein-1 and may be useful for early diagnosis. Clin Exp Immunol, 2011. 164(1): p. 57-65.

[148] Shabo, I. and J. Svanvik, Expression of macrophage antigens by tumor cells. Adv Exp Med Biol, 2011. 714: p. 141-50.

[149] Koch, M., et al., CD36-mediated activation of endothelial cell apoptosis by an N-terminal recombinant fragment of thrombospondin-2 inhibits breast cancer growth and metastasis in vivo. Breast Cancer Res Treat, 2011. 128(2): p. 337-46.

[150] Maiti, S.N., et al., Beta-2-glycoprotein 1-dependent macrophage uptake of apoptotic cells. Binding to lipoprotein receptor-related protein receptor family members. J Biol Chem, 2008. 283(7): p. 3761-6.

[151] Vandivier, R.W., et al., Role of surfactant proteins A, D, and C1q in the clearance of apoptotic cells in vivo and in vitro: calreticulin and CD91 as a common collectin receptor complex. J Immunol, 2002. 169(7): p. 3978-86.
[152] Ghosal, D., N.W. Shappell, and T.W. Keenan, Endoplasmic reticulum lumenal proteins of rat mammary gland. Potential involvement in lipid droplet assembly during lactation. Biochim Biophys Acta, 1994. 1200(2): p. 175-81.
[153] Li, Y., et al., In vitro invasiveness of human breast cancer cells is promoted by low density lipoprotein receptor-related protein. Invasion Metastasis, 1998. 18(5-6): p. 240-51.
[154] Rogers, N.J., et al., A defect in Marco expression contributes to systemic lupus erythematosus development via failure to clear apoptotic cells. J Immunol, 2009. 182(4): p. 1982-90.
[155] Lofgren, K.A., et al., Mammary gland specific expression of Brk/PTK6 promotes delayed involution and tumor formation associated with activation of p38 MAPK. Breast Cancer Res, 2011. 13(5): p. R89.
[156] Cheng, X., et al., Activation of murine double minute 2 by Akt in mammary epithelium delays mammary involution and accelerates mammary tumorigenesis. Cancer Res, 2010. 70(19): p. 7684-9.
[157] Dillon, R.L. and W.J. Muller, Distinct biological roles for the akt family in mammary tumor progression. Cancer Res, 2010. 70(11): p. 4260-4.
[158] O'Brien, J. and P. Schedin, Macrophages in breast cancer: do involution macrophages account for the poor prognosis of pregnancy-associated breast cancer? J Mammary Gland Biol Neoplasia, 2009. 14(2): p. 145-57.
[159] Vafaizadeh, V., P.A. Klemmt, and B. Groner, Stat5 assumes distinct functions in mammary gland development and mammary tumor formation. Front Biosci, 2012. 17: p. 1232-50.
[160] Barbieri, I., et al., Stat3 is required for anchorage-independent growth and metastasis but not for mammary tumor development downstream of the ErbB-2 oncogene. Mol Carcinog, 2010. 49(2): p. 114-20.
[161] Connelly, L., et al., Inhibition of NF-kappa B activity in mammary epithelium increases tumor latency and decreases tumor burden. Oncogene, 2011. 30(12): p. 1402-12.
[162] Desrivieres, S., et al., The biological functions of the versatile transcription factors STAT3 and STAT5 and new strategies for their targeted inhibition. J Mammary Gland Biol Neoplasia, 2006. 11(1): p. 75-87.
[163] Rice, M.S., et al., Insulin-like growth factor-1, insulin-like growth factor binding protein-3 and lobule type in the Nurses' Health Study II. Breast Cancer Res, 2012. 14(2): p. R44.
[164] Ghosh, K., et al., Independent association of lobular involution and mammographic breast density with breast cancer risk. J Natl Cancer Inst, 2010. 102(22): p. 1716-23.

[165] Gierach, G.L., L.A. Brinton, and M.E. Sherman, Lobular involution, mammographic density, and breast cancer risk: visualizing the future? J Natl Cancer Inst, 2010. 102(22): p. 1685-7.

[166] Ginsburg, O.M., L.J. Martin, and N.F. Boyd, Mammographic density, lobular involution, and risk of breast cancer. Br J Cancer, 2008. 99(9): p. 1369-74.

[167] Henson, P.M., Dampening inflammation. Nat Immunol, 2005. 6(12): p. 1179-81.

[168] Monks, J., et al., Epithelial cells as phagocytes: apoptotic epithelial cells are engulfed by mammary alveolar epithelial cells and repress inflammatory mediator release. Cell Death Differ, 2005. 12(2): p. 107-14.

[169] Nagaosa, K., A. Shiratsuchi, and Y. Nakanishi, Concomitant induction of apoptosis and expression of monocyte chemoattractant protein-1 in cultured rat luteal cells by nuclear factor-kappaB and oxidative stress. Dev Growth Differ, 2003. 45(4): p. 351-9.

[170] Graham, K., et al., Gene expression profiles of estrogen receptor-positive and estrogen receptor-negative breast cancers are detectable in histologically normal breast epithelium. Clin Cancer Res, 2011. 17(2): p. 236-46.

[171] Truman, L.A., et al., CX3CL1/fractalkine is released from apoptotic lymphocytes to stimulate macrophage chemotaxis. Blood, 2008. 112(13): p. 5026-36.

[172] Fadok, V.A., et al., Exposure of phosphatidylserine on the surface of apoptotic lymphocytes triggers specific recognition and removal by macrophages. J Immunol, 1992. 148(7): p. 2207-16.

[173] Borisenko, G.G., et al., Macrophage recognition of externalized phosphatidylserine and phagocytosis of apoptotic Jurkat cells--existence of a threshold. Arch Biochem Biophys, 2003. 413(1): p. 41-52.

[174] Martin, S.J., et al., Early redistribution of plasma membrane phosphatidylserine is a general feature of apoptosis regardless of the initiating stimulus: inhibition by overexpression of Bcl-2 and Abl. J Exp Med, 1995. 182(5): p. 1545-56.

[175] Suzuki, J., et al., Calcium-dependent phospholipid scrambling by TMEM16F. Nature, 2010. 468(7325): p. 834-8.

[176] Hanayama, R., et al., Identification of a factor that links apoptotic cells to phagocytes. Nature, 2002. 417(6885): p. 182-7.

[177] Hanayama, R., et al., Autoimmune disease and impaired uptake of apoptotic cells in MFG-E8-deficient mice. Science, 2004. 304(5674): p. 1147-50.

[178] Nakano, T., et al., Cell adhesion to phosphatidylserine mediated by a product of growth arrest-specific gene 6. J Biol Chem, 1997. 272(47): p. 29411-4.

[179] Ishimoto, Y., et al., Promotion of the uptake of PS liposomes and apoptotic cells by a product of growth arrest-specific gene, gas6. J Biochem, 2000. 127(3): p. 411-7.

[180] Hall, M.O., et al., Both protein S and Gas6 stimulate outer segment phagocytosis by cultured rat retinal pigment epithelial cells. Exp Eye Res, 2005. 81(5): p. 581-91.

[181] Reddien, P.W. and H.R. Horvitz, The engulfment process of programmed cell death in caenorhabditis elegans. Annu Rev Cell Dev Biol, 2004. 20: p. 193-221.

[182] Ravichandran, K.S. and U. Lorenz, Engulfment of apoptotic cells: signals for a good meal. Nat Rev Immunol, 2007. 7(12): p. 964-74.
[183] Reddien, P.W. and H.R. Horvitz, CED-2/CrkII and CED-10/Rac control phagocytosis and cell migration in Caenorhabditis elegans. Nat Cell Biol, 2000. 2(3): p. 131-6.
[184] Gumienny, T.L., et al., CED-12/ELMO, a novel member of the CrkII/Dock180/Rac pathway, is required for phagocytosis and cell migration. Cell, 2001. 107(1): p. 27-41.
[185] Hasegawa, H., et al., DOCK180, a major CRK-binding protein, alters cell morphology upon translocation to the cell membrane. Mol Cell Biol, 1996. 16(4): p. 1770-6.
[186] Kiyokawa, E., et al., Activation of Rac1 by a Crk SH3-binding protein, DOCK180. Genes Dev, 1998. 12(21): p. 3331-6.
[187] Van Aelst, L. and C. D'Souza-Schorey, Rho GTPases and signaling networks. Genes Dev, 1997. 11(18): p. 2295-322.
[188] Nakaya, M., et al., Opposite effects of rho family GTPases on engulfment of apoptotic cells by macrophages. J Biol Chem, 2006. 281(13): p. 8836-42.
[189] Riento, K. and A.J. Ridley, Rocks: multifunctional kinases in cell behaviour. Nat Rev Mol Cell Biol, 2003. 4(6): p. 446-56.
[190] Wang, Q.Q., et al., Integrin beta 1 regulates phagosome maturation in macrophages through Rac expression. J Immunol, 2008. 180(4): p. 2419-28.
[191] Stephens, L., C. Ellson, and P. Hawkins, Roles of PI3Ks in leukocyte chemotaxis and phagocytosis. Curr Opin Cell Biol, 2002. 14(2): p. 203-13.
[192] Lerm, M., et al., Inactivation of Cdc42 is necessary for depolymerization of phagosomal F-actin and subsequent phagosomal maturation. J Immunol, 2007. 178(11): p. 7357-65.
[193] Erwig, L.P., et al., Differential regulation of phagosome maturation in macrophages and dendritic cells mediated by Rho GTPases and ezrin-radixin-moesin (ERM) proteins. Proc Natl Acad Sci U S A, 2006. 103(34): p. 12825-30.
[194] Oshima, K., et al., Secretion of a peripheral membrane protein, MFG-E8, as a complex with membrane vesicles. Eur J Biochem, 2002. 269(4): p. 1209-18.
[195] Hanayama, R. and S. Nagata, Impaired involution of mammary glands in the absence of milk fat globule EGF factor 8. Proc Natl Acad Sci U S A, 2005. 102(46): p. 16886-91.
[196] Sanchez-Mejorada, G. and C. Rosales, Signal transduction by immunoglobulin Fc receptors. J Leukoc Biol, 1998. 63(5): p. 521-33.
[197] Silva, M.T., Secondary necrosis: the natural outcome of the complete apoptotic program. FEBS Lett, 2010. 584(22): p. 4491-9.
[198] Stein, T., et al., A mouse mammary gland involution mRNA signature identifies biological pathways potentially associated with breast cancer metastasis. J Mammary Gland Biol Neoplasia, 2009. 14(2): p. 99-116.
[199] Deonarine, K., et al., Gene expression profiling of cutaneous wound healing. J Transl Med, 2007. 5: p. 11.
[200] McDaniel, S.M., et al., Remodeling of the mammary microenvironment after lactation promotes breast tumor cell metastasis. Am J Pathol, 2006. 168(2): p. 608-20.

[201] Schedin, P., Pregnancy-associated breast cancer and metastasis. Nat Rev Cancer, 2006. 6(4): p. 281-91.

[202] Schedin, P., et al., Microenvironment of the involuting mammary gland mediates mammary cancer progression. J Mammary Gland Biol Neoplasia, 2007. 12(1): p. 71-82.

[203] Aziz, M., et al., Review: milk fat globule-EGF factor 8 expression, function and plausible signal transduction in resolving inflammation. Apoptosis, 2011. 16(11): p. 1077-86.

[204] Linger, R.M., et al., TAM receptor tyrosine kinases: biologic functions, signaling, and potential therapeutic targeting in human cancer. Adv Cancer Res, 2008. 100: p. 35-83.

[205] Frisdal, E., et al., Interleukin-6 protects human macrophages from cellular cholesterol accumulation and attenuates the proinflammatory response. J Biol Chem, 2011. 286(35): p. 30926-36.

[206] Png, K.J., et al., A microRNA regulon that mediates endothelial recruitment and metastasis by cancer cells. Nature, 2012. 481(7380): p. 190-4.

Chapter 2

Apoptosis and Activation-Induced Cell Death

Joaquín H. Patarroyo S. and Marlene I. Vargas V.

Additional information is available at the end of the chapter

1. Introduction

In 1885, Flemming reported the first morphological description of the natural process of cell death. This process is now known as apoptosis, a name that was chosen by Kerr [1] to describe the unique morphology associated with this type of cellular death, which is different from necrosis. After the initial description of apoptosis, it was recognized that this process occurs in all tissues as part of the normal cellular turnover. Apoptosis also occurs during embryogenesis, in which particular cells are 'programmed' to die, and hence the term 'programmed cell death' is used to describe this process.

Currently, cell death can be classified according to the morphological appearance of the lethal process (apoptotic, necrotic, autophagic or associated with mitosis), the enzymological criteria (with or without the involvement of nucleases or distinct classes of proteases, such as caspases or cathepsins), the functional aspects (programmed or accidental, physiological or pathological) or the immunological characteristics (immunogenic or non-immunogenic) [2,3].

Apoptosis is a genetically predetermined mechanism that may be elicited through a number of molecular pathways, the best characterized and most prominent of which are called the extrinsic and intrinsic pathways (Fig.1).

In the extrinsic pathway, which is also known as the "death receptor pathway", apoptosis is triggered by the ligand-induced activation of death receptors at the cell surface, which include the tumor necrosis factor (TNF) receptor-1, CD95/Fas (the receptor of CD95L/FasL), and the TNF-related apoptosis inducing ligand receptors-1 and -2 (TRAIL-R1/2). In the intrinsic pathway, which is also called the "mitochondrial pathway", apoptosis results from an intracellular cascade of events in which mitochondrial permeabilization plays a crucial role [4].

In vitro studies showed evidence that all animal cells constitutively express the proteins needed to undergo apoptosis [5,6]. Apoptosis is a process that involves a variety of signaling pathways that lead to multiple cellular changes throughout the cell death process [7,8].

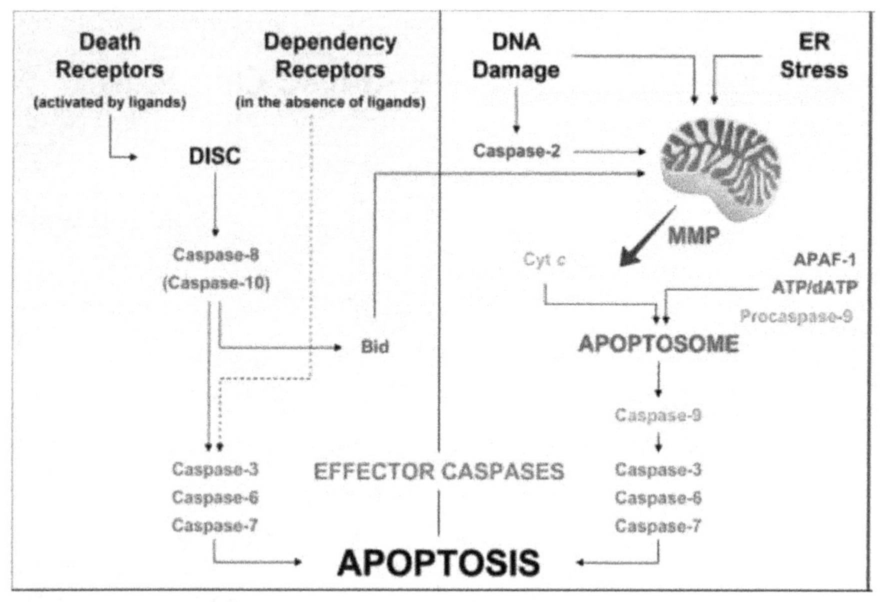

Figure 1. Extrinsic versus intrinsic caspase activation cascades. *Left:* extrinsic pathway. The ligand-induced activation of death receptors induces the assembly of the death-inducing signaling complex (DISC) on the cytoplasm side of the plasma membrane. This promotes the activation of caspase-8 (and possibly of caspase-10), which in turn is able to cleave effector caspase-3, -6, and -7. Caspase-8 can also proteolytically activate Bid, which promotes mitochondrial membrane permeabilization (MMP) and represents the main link between the extrinsic and intrinsic apoptotic pathways. The extrinsic pathway includes also the dependency receptors, which deliver a death signal in the absence of their ligands, through yet unidentified mediators. *Right:* intrinsic pathway. Several intracellular signals, including DNA damage and endoplasmic reticulum (ER) stress, converge on mitochondria to induce MMP, which causes the release of proapoptotic factors from the intermembrane space (IMS). Among these, cytochrome c (Cytc) induces the apoptosis protease-activating factor 1 (APAF-1) and ATP/dATP to assemble the apoptosome, a molecular platform which promotes the proteolytic maturation of caspase-9. Active caspase-9, in turn, cleaves and activates the effector caspases, which finally lead to the apoptotic phenotype. DNA damage may signal also through the activation of caspase-2, which acts upstream mitochondria to favor MMP. Kroemer G, Galluzzi L, Brenner C. (2007).

The process of apoptosis has been conserved throughout evolution. Because of its conserved and uniform nature [8], apoptosis is frequently defined mechanistically as a pathway of regulated cell death that involves the sequential activation of caspases [Cysteine ASPartate-Specific ProteASEs], which are the major effectors of apoptosis [9]. Many caspases are present in healthy cells as catalytically dormant pro-enzymes [zymogens] that have very low enzymatic activity [10] but become activated at the onset of apoptosis through activation signals [11]. The proteolytic cascade in which one caspase can activate other caspases amplifies the apoptotic signaling pathway and thus leads to rapid cell death. However, not all caspases are required for apoptosis. In fact, the process generally requires the activation of a limited subset of caspases; in particular, caspases-3, -6, and -7 are the "executioner" caspases and these mediate their effects through the cleavage of specific substrates in the cell [12]. Over the last 10 years, intense research has focused on the pathways that control caspase activation. Some of these pathways, such as the apoptosome and death receptor-mediated pathways and the signalosomes responsible for caspase activation, are well established, whereas others are less clearly defined [12-14].

Other proteases, such as granzymes and calpain, are also involved in the apoptotic signaling process, but in a much more cell type- and/or stimulus-specific manner. At least three distinct caspase-signaling pathways exist: one is activated through the ligand-dependent death receptor oligomerization, the second through mitochondrial permeabilization, which leads to the release of proapoptotic proteins from the mitochondrial intermembrane space [15], and the third through stress-mediated events that involve the endoplasmic reticulum [16,17]. These pathways can also interact to amplify weak apoptotic signals. Some studies about programmed cell death indicate that apoptosis may occur in the complete absence of caspases; these instances include when organelles, such as the mitochondria, endoplasmic reticulum or lysosomes, are damaged leading to an increased release of calcium, species of oxygen free radicals and effectors proteins [18-20].

Formation of the death-inducing signaling complex (DISC) or the apoptosome activates initiator and common effectors caspases that execute the apoptosis process; both of these types of caspases are found in the cytoplasm. However, there is now evidence that the two pathways are linked and that molecules from one pathway can influence the other [16-24]. In addition, another mechanism activates a parallel caspase-independent cell death pathway via single stranded DNA damage [25].

2. Apoptosis mechanisms and pathways

Studies from the 1970s and 1980s have shown that apoptosis can be defined by specific morphological characteristics and is regulated by specific biochemical processes. Studies using the nematode *Caenorhabditis elegans* have contributed most of the current understanding of the apoptosis mechanisms [26,27]. The final phase of apoptosis is the activation of endonucleases, which leads to the fragmentation of DNA. Apoptosis is genetically predetermined and there are therefore genes that can promote or inhibit it [27]; these genes can respond to both normal and pathological stimuli.

Individual genes have been associated with apoptosis in two ways: the genes either are expressed in cells undergoing apoptosis or modulate the effects of the process. Among these, the proto-oncogene c-myc triggers either cell proliferation or apoptosis. Under normal conditions, this gene programs the cell to grow; however, if this procedure is prevented due to the lack of growth factors or the presence of secondary oncogenes, the cell enters the death process. The expression of c-myc is required for this process most likely because c-myc influences the delicate balance between the survival and death signaling pathways that are simultaneously activated by this factor [28].

The caspases are directly or indirectly responsible for the morphological and biochemical changes that characterize the process of apoptosis [29]. The machinery of cell death effectors that is managed by the family of caspases cleaves many vital proteins and proteolytically active enzymes, a process that contributes to the destruction of the cell. All caspases are initially synthesized as inactive zymogens that contain a prodomain that is formed by a p20 large subunit and a p10 small subunit. The activation of this zymogen precursor is mediated by a series of cleavage events, especially for the executioner caspases, which first separate the large and small subunits and then remove the prodomain. The active enzyme is composed of a heterotetramer that is formed by two large and two small subunits [30].

To date, 14 mammalian caspases have been identified. These can be subdivided into families based on their sequence homology and substrate specificity [31]. Caspases can be classified as: Caspase-1 (ICE), caspase-2 (ICH-l, Nedd-2), caspase-3 (CPP32, Apopain, Yama), caspase-4 (ICH-2, TX, ICEre), caspase-5 (ICErel, TY), caspase-6 (Mch2), caspase-7 (ICE-LAP3, Mch3, CMH-1), caspase-8 (FLICE,Mch5, MACH), caspase-9 (Mch6, ICELAP6), caspase-10 (Mch4), caspase-11 (ICH-3), caspase-12, caspase-13 (ERICE) and caspase-14 (MICE) [6]. Currently, 11 human caspases have been identified: caspase-1 through -10 and caspase-14 [9,32,33]. The protein initially named caspase-13 was later found to represent a bovine homologue of caspase-4 [34] and caspase-11 and -12 are murine enzymes that are most likely the homologues of the human caspase-4 and -5. However, only some of these caspases have been found to be involved in the process of apoptosis; these are caspase-2, caspase-3, caspase-6, caspase-7, caspase-8, caspase-9, caspase-10 and caspase-12 [35].

Caspases can also be classified as either initiators or effectors. The upstream (initiator) caspases, unlike the downstream (effector) caspases, have long prodomains with structural motifs (e.g., death effector domain, DED, or caspase recruitment domain, CARD) that associate with their specific activators [36]. Any apoptotic extracellular or intracellular signal is transduced by adapter proteins and transmitted to specific cysteine proteases called "initiator caspases", which commit the cell to apoptosis. However, the activated caspases can be inhibited by endogenous inhibitors, such as IAPs (inhibitors of apoptosis). The commitment is followed by the "execution" of the cells through the sequential activation of the "executioner caspases" and the systematic disintegration of the cellular structure, which is followed by the disposal of the dead cells by phagocytosis. The adapter proteins include molecules that are homologous to the nematode protein CED-4, whereas initiator and executioner caspases are homologues of CED-3. Each of the initiator caspases and the respective adapter molecules, which are responsible for the oligomerization of the caspase to which it binds, define distinct pathways for apoptotic caspase activation and cell death [37].

The apoptotic signaling pathways that lead to zymogen processing can be subdivided into two major categories: cell surface or intracellular sensor-mediated. The former pathway is activated in response to extracellular signals, which indicate that the existence of the cell is no longer needed for the well-being of the organism. These cell surface sensor-mediated apoptotic signals are initiated by the binding of ligands to cell surface death-mediating receptors, which include the death receptor family [38].

During the past two decades, extensive work has been performed to elucidate the molecular mechanism of apoptosis. It is clear that apoptosis is induced by a range of stimuli that activate two major cell death signaling pathways: the intrinsic pathway, which is mainly controlled by the Bcl-2 protein family members, and the extrinsic pathway, which is activated by the death receptors of the tumor necrosis factor receptor superfamily.

The extrinsic pathway is triggered by death receptor engagement, which initiates a signaling cascade that is mediated by the activation of caspase-8. Specifically, apoptosis is induced by the interaction of a death receptor, namely, Fas (APO-1, CD95), TNF receptor-1 (TNFR-1), DR-3 (TRAMP), DR-4 (TRAIL-R1) or DR-5 (TRAIL-R2), with its respective ligand [39]. These cell surface receptors, which are located on the cell membrane, are members of the TNFR family. The binding of a ligand to a death receptor cause its oligomerization, which results in its activation. The oligomerization of the receptors is followed by the binding of specific adapter proteins (FADD, TRADD) to the receptor complex, which results in the recruitment of the procaspase-8 and -10 to the receptor complex where the proenzymes become activated. Active caspase-8 and -10, in turn, active through effector caspase-3 the caspase cascade [35]. The best-characterized ligands and their corresponding death receptors include FasL/FasR, TNF-α/TNFR1, Apo3L/DR3, Apo2L/DR4 and Apo2L/DR5 [40-43]. Alternatively, an extrinsic pro-apoptotic signal can be dispatched by the so-called 'dependence receptors', including netrin receptors (e.g., UNC5A-D and deleted in colorectal carcinoma, DCC). These receptors induce apoptosis in the absence of the required stimulus (when unoccupied by a trophic ligand, or possibly when bound by a competing 'antitrophin'), but block apoptosis following binding of their respective ligands [2,3].

The intrinsic pathway is activated when various apoptotic stimuli trigger the release of cytochrome c from the mitochondria, which is independent of caspase-8 activation. This activation occurs through the formation of an "apoptosome", which consists of cytochrome c, the apoptotic protease activating factor-1 (Apaf-1) and procaspase-9. The formation of the apoptosome is dependent on both the release of cytochrome *c* from the mitochondria and free ATP or dATP in the cell [42-45]. Together with ATP/dATP, cytochrome *c* binds to Apaf-1, a cytosolic protein, and induces its oligomerization, which leads to the recruitment of an initiator caspase, procaspase-9, which undergoes a conformational change that results in caspase-9 activation. These proteins (adaptor proteins and procaspases) interact via their caspase recruitment domains (CARD). The apoptosome then recruits procaspase-3, which is subsequently cleaved, and therefore activated, by the active caspase-9 and then released to mediate apoptosis. This mechanism is caspase-dependent; however, a caspase-independent cell death (CICD) pathway also exists. In this mechanism, the apoptosis-induced factor (AIF) and endonuclease G (EndoG) are released from the mitochondria and relocate to the nucleus, where they mediate the large-scale fragmentation of DNA independently of caspases [46].

The mitochondrial outer membrane (MOM) permeabilization is considered the "point of no return" for apoptotic cell death and triggers the release of proteins that mediate cell death, such as cytochrome c, into the cytoplasm [46,47].

The intrinsic cell death signals generally converge within the cell at the MOM and result in the loss of mitochondrial membrane integrity and the subsequent activation of the downstream apoptotic pathways. The release of cytochrome *c* and others proteins from the mitochondria is tightly regulated by pro-apoptotic members of the Bcl-2 family, which are believed to act as ion and/or protein channels or possibly as regulators of such channels [14,15,19]. The anti-apoptotic members of this family protect the cells from apoptosis through the sequestration of pro-apoptotic proteins or by interfering with their activities [48-50].

The DR (extrinsic) and mitochondrial (intrinsic) apoptosis pathways have been described in detail [51-57].

The pathway that is utilized depends on the initial death signal, the cell type involved, and the balance between pro-apoptotic and anti-apoptotic signals [58]. In addition, the initiation of a specific pathway may exhibit cross-talk with another, which might result in the activation of a second pathway. Although there is a wide variety of physiological and pathological stimuli and conditions that can trigger apoptosis, not all cells will necessarily die in response to the same stimulus.

Recent studies implicate the ER as a third sub-cellular compartment that can signal apoptosis via a signaling pathway that is called the ER-stress pathway [59]. Prolonged ER stress or the mobilization of intracellular Ca^{++} stores stimulates the cleavage and activation of pro-caspase-12 (in mice), which is localized in the ER membrane, by m calpain. Once activated, caspase-12 then activates executioner caspases to induce apoptosis. Studies have shown that prolonged ER stress can result in the activation of caspase-12, which initiates apoptosis [56,60].

Several caspase regulators have been discovered, including activators and inhibitors of these proteases [22]. Inhibitor of apoptosis proteins (IAPs) interfere with apoptosis by blocking caspase activity. Therefore, because of their inhibitory activity against the executioner caspases, IAPs may inhibit all caspase-dependent apoptosis. However, Smac/DIABLO, a mitochondrial protein released in the cytosol in response to an apoptotic stimulus, inhibits the anti-apoptotic function of several members of the IAP family, thereby de-repressing caspase activation [61].

The intrinsic pathway is marked by a requirement for the involvement of mitochondria. Under the control of the BCL-2 family the mitochondria participate in apoptosis by releasing apoptogenic factors.

The Bcl-2 family members can be divided into three subfamilies based on their structural and functional features [62]. The anti-apoptotic subfamily contains the Bcl-2, Bcl-XL, Bcl-w, Mcl-1, Bfl1/A-1, and Bcl-B proteins. The members of this subfamily suppress apoptosis and contain all four Bcl-2 homology (BH) domains. Some pro-apoptotic proteins, such as Bax, Bak, and Bok, contain BH 1-3 domains and are therefore termed "multidomain" whereas other pro-apoptotic proteins possess only BH3 domain and are thus referred to as "BH3-only" proteins [63].

The intrinsic pathway, which is also called the BCL-2-regulated or mitochondrial pathway (in reference to the role this organelle plays), is activated by various developmental cues or cytotoxic stimuli such as viral infection, DNA damage and growth factor deprivation, and is strictly controlled by the BCL-2 family of proteins [64]. This pathway predominantly leads to the activation of caspase-9 [65], although, at least in certain cell types, it can proceed in the absence of caspase-9 or its activator, the apoptotic protease-activating factor-1 (Apaf1) [11]. The extrinsic, or DR pathway, is triggered by the ligation of the so-called death receptors, such as Fas or TNF-R1, which contain an intracellular death domain that associates with adaptor proteins that can recruit and activate caspase-8 through their Fas-associates death domains (FADD; also known as MORT1). This activation of caspase-8 causes the subsequent activation of downstream caspases, such as caspase-3, -6 or -7, and does not necessarily involve the BCL-2 protein family [64,66].

3. Physiological cell death in the immune system

In the immune system, the death of T and B lymphocytes induced by specific receptors may occur in the central and peripheral lymphoid compartments. Apoptosis is the most common form of cell death in the immune system [67].

Apoptosis occurs in the primary lymphoid organs, such as the bone marrow, the liver and the thymus, and is used to eliminate useless precursor cells with non-rearranged, or aberrantly rearranged, non-functional receptors. Furthermore, apoptosis is essential for the deletion of autoreactive T cells in the thymus and is therefore crucial in the assurance of central self-tolerance [68,69].

A similar apoptotic deletion mechanism operates in the T and B cells that are located in the peripheral lymphoid organs, such as the lymph nodes and the spleen. This cell deletion by apoptosis is another safeguard of the immune system to assure self-tolerance and to downregulate an excessive immune response. Lymphocytes that escape this process probably replenish the pool of cells that determine immunological memory [70-73].

4. Antigenic stimulation drives proliferation and death

The immune system has, among other characteristics, its specificity: the repertoire of T and B lymphocytes, initially built from randomly selected antibody and T cell receptor (TCR) variable region genes, is shaped by selection to cope, on the one hand, with the vast universe of antigens and, on the other hand, with the danger of autoimmunity [74]. Another distinctive feature is the control of homeostasis; after a clonal expansion phase, the antigen-reactive lymphocytes must be deleted until the pool of lymphoid cells reaches its baseline level. This controlled cell death is achieved by a fine-tuned balance between growth/expansion and apoptotic death; in general, because the immune system produces more cells than it needs, the extra cells are eliminated by apoptosis [75].

The thymocytes that do not express a functional TCR die due to the lack of survival signals, an event that is known as "death by neglect". In contrast, thymocytes bearing a TCR that recognizes self-peptides that are presented by major histocompatibility complex (MHC) molecules are eliminated by apoptosis in response to a high-affinity signal; this process is termed negative selection [76,77]. Moreover, the normal immune response requires the regulated elimination of specific cell populations by apoptosis. During the development of T cells in the thymus and B cells in the bone marrow, all potentially autoreactive lymphocytes are removed by apoptosis. Therefore, the initial overproduction is followed by the death of those cells that fail to exhibit productive antigen specificities [49]. In addition, after the inciting antigen in an immune response has been cleared, cell death mechanisms return the number of lymphocytes to its normal physiological range.

Only those lymphocytes bearing an antigen receptor with an appropriate specificity are selected for survival and further differentiation; the remaining (~75% and ~95% of B-cell and T-cell precursors, respectively) undergo apoptosis [78-81].

The immature precursor T lymphocytes develop into mature antigen-reactive T cells in the thymus. The maturation process is associated with the acquisition of high expression levels of the TCR-CD3 complex and with tolerance to self antigens; these two characteristics are the consequence of positive and negative selection, respectively. Positive selection ensures the survival of immature T cells ($CD4^+CD8^+$) that have an appropriate TCR [53,82].

The number of T cells that leave the thymus and enter the peripheral T-cell pool is approximately 2–3% of the number of cells that were initially generated. The pre-T lymphocytes in the thymus undergo differentiation and TCR rearrangement. Those T cells that fail to rearrange their TCR genes productively and thus cannot be stimulated by self-MHC–peptide complexes die by neglect. In contrast, cells with autoreactive receptors are killed by activation-induced death [83].

Those thymocytes that successfully pass the pre-TCR selection develop into $CD4^+CD8^+$ T cells and undergo further TCR-affinity-driven positive and negative selection. After these selection processes, the mature single-positive CD4+ MHC class-II restricted and CD8+ MHC class-I-restricted T cells leave the thymus and generate the peripheral T-cell pool [52].

There exists evidence for the positive selection of B cells based on B cell antigen receptor [BCR] specificity but the nature of the selecting ligands is not yet known. The processes of antigen receptor gene rearrangement and diversification can produce self-antigen specific B and T lymphoid cells that could initiate autoimmune tissue destruction [11].

Apoptosis also occurs as a defense mechanism in immune reactions or when cells are damaged by disease or noxious agents [84]. During infection, the lymphocytes and other cells of the innate immune system, which express receptors that recognize foreign antigens, undergo proliferation and differentiation to develop effectors functions that help kill the invading pathogens. These effectors functions, which include cellular- or antibody-mediated cytotoxicity and inflammatory cytokines, can be harmful to the host. Therefore, to limit the damage to healthy tissue, mechanisms, such as cell inactivation and cell death, have evolved to shut down these immune responses [85].

5. Activation-induced cell death

The term "activation-induced cell death" (AICD) describes the signal-induced programmed cell death of T lymphocytes [86] and distinguishes this phenomenon, or apoptosis process, from other possible effects of antigen receptor ligation (e.g., cytokine production and clonal expansion). However, apoptosis via AICD is almost certainly a major mechanism of clonal deletion in the immune system.

The AICD process is induced by the same signals that, in other circumstances, can lead to T cell proliferation and activation; these signals include antigen recognition, the binding of an antibody anti CD3 and the exposure to mitogens [87]. The AICD may occur in immature or transformed T cells, as well as in mature and activated peripheral blood T cells [88-94].

The primary function of T lymphocytes is to mount an efficient immune response to an antigen. However, the maintenance of T lymphocyte homeostasis between antigenic challenges is also essential. Co-stimulatory molecules are required for efficient T cell responses and an IL-7 signal is essential for maintaining the homeostasis of both naïve and memory T cell populations. Both the intrinsic and the extrinsic apoptotic pathways are actively involved in the regulation of T cell responses and homeostasis. Several anti-apoptotic molecules have been identified as critical effectors molecules in mature T lymphocytes and the co-stimulatory molecules signaling pathways [95].

The proliferation and differentiation of T- and B-cells following antigen stimulation are crucial in the adaptive immune response. In fact, the activation of T cells is a critical step that occurs early in the adaptive immune response.

The TCR is a transmembrane protein that consists of a heterodimer, which can be one of two types. In 95% of T cells, the heterodimer consists of an alpha (α) and a beta (β) chain, which are expressed on the surface of T lymphocytes as part of a multi-subunit complex with the CD3 protein; in 5% of T cells, however, the heterodimer consists of gamma and delta (γ/δ) chains. When the TCR engages with the antigen that is presented on the MHC, the T lymphocyte is activated through a series of biochemical events that are mediated by a number of enzymes, co-receptors, accessory molecules, and activated or released transcription factors. The $\alpha\beta$ chains of the TCR are joined by disulfide bonds and have a structure similar to immunoglobulins with an extracellular domain, which is responsible for the recognition of the antigenic peptide, a transmembrane domain and a short intracytoplasmatic domain [96,97].

The induction of apoptosis via signaling through DR requires the close proximity of the intracytoplasmatic or transmembrane domain of several receptor molecules because their activation requires the cross-linking of several receptors [98-100].

Antigen-presenting cells (APCs) are specialized white blood cells that help fight off foreign substances that enter the body. Enzymes inside the APCs break down the antigen into smaller particles. The processed antigens are transported to the surface of the APCs, bound with either an MHC class I or class II molecule. This complex forms epitopes (part of a foreign substance that can be recognized by the immune system), which the TCR recognizes and binds to.

Two signals derived from the APCs are required for the activation-induced proliferation of T cells. The first signal is the engagement of the TCR to the MHC-antigen complex on the APC and the second involves the coupling of co-stimulatory molecules, such as CD28, on the T cell with the B7 protein family expressed by the APCs. The signaling pathways downstream of the TCR and CD28 involve the integration of complex signals that lead to activation, proliferation, cell survival, and death. These signals are all mediated by cell surface receptor-ligand interactions and cytokines, such as interleukin IL-2, IL-4, and IL-10, which are produced by accessory cells as well as the responding T cells themselves [101].

The CD3 protein complex, which is composed of five polypeptides (γ δ, ε, ζ, η), is non-covalently associated with the TCR. Of the polypeptides, 90% the ζ chains form homodimers and 10% heterodimers with the η chains. The remaining three polypeptides associate to form the heterodimers $\varepsilon\delta$ and $\varepsilon\gamma$, which are also transmembrane proteins with an intracytoplasmatic domain that is larger than the TCR $\alpha\beta$ and is responsible for translating the activation signals through the complex to the inside of the cell [102]. The TCR, the ζ-chain, and the CD3 molecules comprise the TCR complex. The *in vitro* stimulation of T cells with anti-CD3 antibody induces a signaling response that stimulates the activity of protein tyrosine kinase (PTK) within seconds [103]. The first event leads to a series of biochemical exchanges that occur over a period of many hours; these include the expression of cytokine receptors, the secretion of cytokines, the initiation of DNA replication and the acquisition of a differentiated phenotype [7,104].

The intracellular tails of the CD3 molecules contain a single conserved motif known as the immunoreceptor tyrosine-based activation motif (ITAM), which is essential for the signaling capability of the TCR. The PTK activity that is induced by the binding of a ligand to the TCR initiates the phosphorylation of tyrosine residues on a number of soluble and membrane-associated substrates. The activation of the ITAMs causes it to act as the substrate for PTK and recruits this protein to the TCR [105,106].

TCR engagement by antigens triggers the tyrosine phosphorylation of the ITAMs, present in the TCR-CD3 complex. Such ITAMs function by orchestrating the sequential activation of the Src-related PTKs: Lck and Fyn, which initiate TCR signaling, followed by that of ZAP70, which further amplifies the response. Lck is activated by the interaction of MHC-II and CD4 or CD8. These various PTKs induce tyrosine phosphorylation of several polypeptides, including the transmembrane adaptor linker activator for T-Cells (LAT). Protein tyrosine phosphorylation subsequently leads to the activation of multiple pathways, including extracellular signal regulated kinase (ERK), c-Jun N-terminal kinase (JNK), nuclear factor kappa B and nuclear factor of activated T-Cells (NF-κB and NFAT) pathways, which ultimately induce effectors functions [105].

The AICD in peripheral T cells depends on the activation state of the cell, i.e., recently activated T cells are resistant to the induction of apoptosis and this initial resistance is followed by a subsequent sensitive phenotype.

The AICD occurs by either a suicidal or fratricidal mechanism through the activation, after TCR stimulation, of either Fas or TNF-R2, respectively. The importance of the Fas/FasL system for the activation of AICD was demonstrated by the discovery of a soluble receptor called decoy receptor 3 (DcR3) which is a member of TNFR superfamily. It has been shown to be the decoy receptor for FasL, which inhibits FasL-mediated apoptosis [75].

Expression of FasL is restricted to the lymphoid organs, and defects in its expression are associated with pathophysiological and autoimmune disorders. The consequences of FasL shedding are quite substantial because only membrane-bound FasL (mFasL) triggers the death signal [106].

The TNF receptor family can also transduce signals via intermediary molecules that can lead, in some instances, to apoptosis via the activation of FADD and FLICE, which also mediate death signals from CD95 [54,108-112].

The first role described for the FADD protein was its interaction with the membrane-bound DRs, which leads to the hypothesis that it is cytoplasmic and only implicated in death signaling pathways. However, new evidence shows that FADD is a much more complex molecule. Depending on its phosphorylation state and subcellular localization, the FADD protein can induce apoptosis, survival, cell cycle progression or T cell proliferation [113].

During development and maturation, lymphocytes with antigen receptors that are not properly rearranged, not positively selected or potentially self-reactive die by apoptosis. Similarly, immature cells that are not selected to enter the pool of mature circulating cells and peripheral cells that exert reactivity for self-antigens undergo apoptosis [92].

Several cells types are involved in the immune response. First, the B and T lymphocytes recognize antigen. After antigen binding to their antigen-specific receptors, these cells undergo clonal expansion and differentiation into effectors cells [114].

The developing lymphocytes express antigen receptors of random specificity. If, within a given window of time, the antigen receptors in the lymphocyte are ligated, the cell is deleted, which might occur through elimination or functional deletion by permanent unresponsiveness. If this ligation does not occur, the cell may mature and subsequently bind to the antigenic stimulus, thereby driving the immune response. The process of clonal deletion in the regulation of central tolerance is now well established [50]. The first unambiguous demonstrations of this process occurred with observations of negative selection, in which developing T lymphocytes in the thymus that expressed antigen receptors disappeared from the T-cell pool.

The regulation of the life and death of T cells uses molecular mechanisms that may be different in distinct T-cell populations depending on their activation state [115].

6. Dynamics of the immune response in peripheral lymphoid organs

The induction of immunity occurs almost exclusively in the secondary lymphoid organs. Consequently, the examination of the lymphoid tissues allowed the elucidation of many of the steps that are involved in the generation of immunity.

In secondary lymphatic tissues, which are the primary locations for the generation of the humoral immune response, the coordinated trafficking of the immune cells is maintained through the direct interaction of the APCs with the T and B cells, which lead to the germinal center formation, the immunoglobulin class switching and the affinity maturation processes of the antibody response [116]. The antigens arrive at the lymph nodes by way of the lymph as soluble antigens, immune complexes or in association with dendritic cells (DCs). The systemic antigens and immune complexes reach the spleen and are first encountered by B cells, macrophages and DCs in the marginal zone (MZ).

The site of B cell differentiation in response to antigen is the germinal center (GC). These structures, which arise within follicles, are responsible for the formation of long-lived memory via the long-lived circulating memory B cells and the long-lived plasma cells that typically reside in the bone marrow [117]. The GC has specific kinetics of expansion, contraction and ultimately dissolution. There is considerable proliferation and cell death within the GC, in addition to cell emigration from the GC [118].

In addition to soluble antigens transport into follicles, molecules smaller than 70–80 kDa are transported through the enclosed compartment of the reticular network, known as the conduit system, into the lumen of the high endothelial venules (HEV). The conduit system of the reticular network is a highly specialized extracellular matrix consisting of a central core, which is formed by interstitial matrix molecules, such as collagen type I and collagen type III. It is also surrounded by a basement membrane-like structure and is ensheathed by a layer of fibroblastic reticular cells [119].

The antigen presentation to B cells by DCs has been described by several groups and it is possible that outer T zone DCs present conduit-derived antigens to B cells. The conduits are also present within follicles but their contribution to antigen delivery within the follicles is not known.

The deletion of B cells that recognize autoantigens occurs at both the immature B cell stage in the bone marrow and the mature B cell stage in the lymph nodes and the spleen. The critical events of B-cell activation, proliferation, differentiation and apoptosis occur in the GCs of the lymphoid follicles. A significant number of B cells die during passage through the GC and the apoptotic B cells are rapidly cleared from this area [120].

The B cells that arrive at the lymph nodes via HEV in the T-zone might encounter antigen-bearing DCs during their migration. Alternatively, follicular B cells that migrate through the B–T boundary region during their random walk might engage the DCs that are concentrated in this region [117].

The production of clusters of antibody secreting cells (ASC) is an important initial component of the B cell response to antigen. It occurs during both T cell-dependent and T cell-independent responses [121]. Furthermore, it has been determined that, depending on the route of antigen delivery, different types of B cells may be involved in focus formation. For example, particulate antigens that are delivered through the blood will, upon entry into the spleen, specifically activate MZ-B cells [122], whereas antigens that are delivered subcutaneously will activate follicular B cells through a complex process of antigen acquisition and B cell trafficking [123].

Apoptotic B cells are found throughout the GCs. Most of the apoptotic B cells, however, are located in the light zone, an area to which B cells from the dark zone immigrate. The dark zone is the site where most of the proliferation of germinal center B cells occurs [118]. Some of the dark zone B cells, termed centroblasts, express CD77 antigen. The CD77 molecule defines a B lymphocyte maturation pathway, specific for GC, where the cells undergo programmed cell death [124]. The apoptosis of germinal center B cells may be a consequence of a lack of stimulation through their cell surface Ig [125]. This lack of stimulation may lead to either affinity maturation, the process by which B cells produce antibodies with increased affinity for an antigen during an immune response, or to a decreased affinity for the immunizing antigen. Thus, B cells with functional receptors are positively selected and differentiate either towards B cell memory or plasma cells [126,127]. Within the lymphoid follicle, APO1/Fas are highly expressed in the areas where B-cell apoptosis occurs [128]. Bcl-2 was shown to mediate an important survival signal for B cells and is believed to be involved in the maintenance of B-cell memory in GCs. The lack of Bcl-2 expression in areas where B cells die by apoptosis probably enables the cells to undergo apoptosis. On the other hand, overexpression of Bcl-2 in these cells would probably inhibit apoptosis [129,130]. There is evidence that the release of apoptogenic molecules from the mitochondria is not necessary for the induction of this pathway [131].

The CD40 receptor can mediate both pro- and anti-apoptotic signals. The binding of the CD40 ligand to its receptor can inhibit the apoptosis of B cells that are stimulated through their Ig receptors and can stimulate B cell proliferation and isotype switching. However, the same receptor-ligand interaction can also induce the expression of CD95 and therefore enhance the susceptibility of the B cell to CD95-mediated death [132].

In peripheral T cells, the triggering of the TCR has several consequences: first, primary activation of resting T cells via the TCR may lead to proliferation of the T-cell population; second, in the absence of co-stimulatory signals, TCR triggering may cause anergy; and, finally, TCR triggering of previously activated T cells may lead to apoptosis unless the T cells are rescued by additional signals [133-135].

Whereas the activation of naïve T cells results in their clonal expansion and differentiation, the repeated stimulation of previously activated cells with antigen triggers AICD, which can be prevented by blocking the FasL/Fas interaction. In addition, the AICD is associated with an activation-induced surface appearance of FasL, which, in turn, triggers death via Fas in an autocrine or paracrine fashion [53]. Therefore, most T cells become susceptible to the induction of apoptosis during their proliferation. This sensitization of T cells is achieved by the regulation of the extrinsic and the intrinsic apoptotic pathways and is influenced by T-cell-associated IL-2 [136], which, in this context, acts as a sensitizer to AICD apoptosis rather than as a growth factor [137]. The molecular mechanism by which this T cell sensitization is achieved involves the altered expression of genes that encode proteins such as BCL-2 family members, hematopoietic progenitor kinase 1 (HPK1), cellular FLICE-like inhibitory protein (cFLIP) family members and CD95L [138,139].

The AICD process involves the stimulation of CD95 [53,99,108], TNFR1 [139], TRAILR in 'helpless' memory CD8 T cells [140-144]. Some findings suggest that TNF is involved in the late phase of AICD [41].

In recent years, our research has focused on the development of a vaccine against *Riphicephalus (Boophilus) microplus.* Using the natural host as experimental model, the kinetics of the immune response in two peripheral lymphoid organs, the spleen and lymph nodes were studied. We performed a kinetic analysis of the apoptotic activity that occurs within the GCs of the draining lymph nodes of animals immunized with a synthetic vaccine after a primary and a secondary stimulation (Fig.2). The examination of the draining lymph nodes of the immunized bovines showed that the apoptosis in the GC occurred mainly in the dark zone seven days after immunization [146]. In parallel, we analyzed the *in vitro* immune memory and AICD of $CD4^{+}$ T cells after stimulation with various inducers (mitramicina and cycloheximide) or inhibitors (benzimide riboside and cyclosporine A) and the synthetic vaccine. The AICD induced by the peptide was mediated by a calcineurin-independent pathway. However, we observed the induction of cell death after stimulation of the TCR by the increased expression of Fas, which is characteristic of AICD.

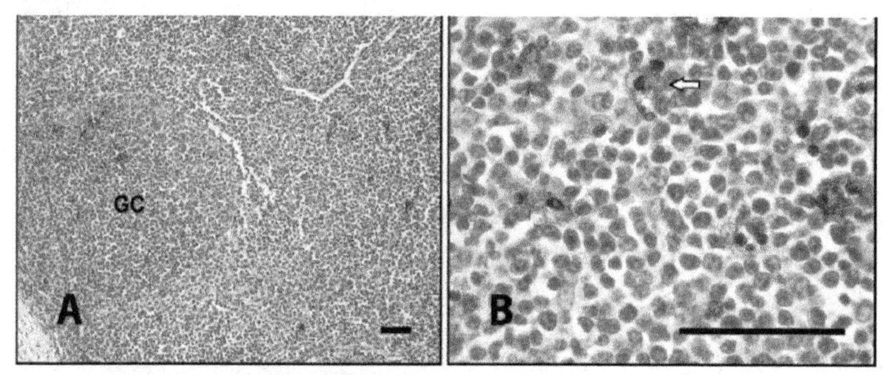

Figure 2. Microphotography of bovine lymph nodes. (A) TUNEL-positive cells in the dark zone of germinal center (GCs). (B) Arrow shows details of TUNEL-positive cells 7 days after immunization. Scale bar 50mm.

7. Concluding remarks

Much progress has been made over the past 15 years analyzing cell death in the immune system. Many cell death regulators have been discovered and their functions defined in experiments with transgenic and knockout mice. In the immune system, apoptosis is seen in cells that are stressed or deprived of growth factors, in developing cells that fail to properly rearrange an antigen receptor or express one that fails negative or positive selection, or in cells that had expanded in response to an antigenic stimulus and are now in excess. We observed that the repeated stimulation of T cells previously activated with synthetic vaccine triggers AICD. Secondary lymphoid organs such as the spleen and lymph nodes provide the critical niches where naïve B cells encounter antigens and produce high-affinity antibodies. Many recent avenues of investigation have elucidated that the germinal center is a dynamic microenvironment where B-cells undergo repeated rounds of mutation and selection. The complex microarchitecture of the germinal center is therefore created not only by distinct stages of B-cell maturation but also by the distribution of immunophenotypically distinct and functionally specialized T, dendritic, and stromal cell subpopulations and their myriad interactions. The GC response is a complex process involving numerous cellular and cell surface components together with multiple signaling pathways. Recent advances in the field include a better understanding of the role of follicular dendritic cells, chemokines and TNF family members in forming and maintaining GC reactions. Even with these new advances, much remains to be understood in GC biology. Not only is there little understanding of the mechanisms that regulate and control GC B cell differentiation, but efforts need to be focused on examining abnormal GC responses, especially in the setting of autoimmunity

Author details

Joaquín H. Patarroyo S.
Laboratory of Biology and Control of Hematozoa and Vectors,
BIOAGRO/Veterinary Department – Federal Viçosa University Viçosa-MG, Brazil

Marlene I. Vargas V.
Laboratory of Immunopathology,
BIOAGRO/Veterinary Department – Federal Viçosa University Viçosa-MG, Brazil

8. References

[1] Kerr JF, Wyllie AH, Currie AR. (1972) Apoptosis: A basic biological phenomenon with wide-ranging implications in tissue kinetics. Brit. J. Cancer, 26:239-257.
[2] Galluzzi L, Maiuri MC, Vitale I, Zischka H, Castedo M, Zitvogel L, Kromer G. (2007) Cell death modalities: classification and pathophysiological implications. Cell Death and Differentiation, 14: 1237–1266.
[3] Galluzzi L, Vitale I, Abrams JM, Alnemri ES, et. al. (2011) Molecular definitions of cell death subroutines: recommendations of the Nomenclature Committee on Cell Death 2012. Cell Death and Differentiation, 1–14.
[4] Degterev A, Junying Y. (2008) Expansion and evolution of cell death programmes. *Cell,* 138(2): 229–232.
[5] Lockshin RA, Williams CM. (1965) Programmed cell death. I. Cytology of degeneration in the intersegmental muscles of the pernyi silkmoth. J. Insect Physiol., 11:123–133.
[6] Lockshin RA, Zakeri Z. (2001) Programmed cell death and apoptosis: origins of the theory. Nat. Rev. Mol. Cell Biol., 2:545–550.

[7] Tamaoki T, Nakano H. (1990) Potent and specific inhibitors of protein kinase C of microbial origin. Bio Technology, 8:732-735.

[8] Cohen JJ, Duke RC (1992) Apoptosis and programmed cell death in immunity. Annu. Rev. Immunol., 10:267-93.

[9] Alnemri ES, Livingston DJ, Nicholson DW, Salvesen G, Thornberry NA, Wong WW, Yuan J. (1996) Human ICE/ CED-3 protease nomenclature. Cell, *87*:171.

[10] Cardone M H, Roy N, Stennicke H R, Salvensen G S, Franke T F, Stanbridge E, Frisch S, Reed JC. (1998) Regulation of cell death protease caspase-9 by phosphorylation. Science, 282:1318-1321.

[11] Crabtree GR (1989) Contingent genetic regulatory events in T lymphocyte activation. Science, 243:355-361.

[12] Kroemer G, Galluzzi L, Brenner C. (2007) Mitochondrial Membrane Permeabilization in Cell Death. Physiol. Rev., 87:99-163.

[13] Strasser A, Bouillet P. (2003) Apoptosis regulation in lymphocyte development, Immunological Reviews, 193:82–92.

[14] Marsden VS, Strasser A (2003) Control of Apoptosis in the Immune System: Bcl-2, BH3-Only Proteins and More. Annu. Rev. Immunol., 21:71–105.

[15] Tait SWG, Green DR. (2010) Mitochondria and cell death: outer membrane permeabilization and beyond. Nature Rev., 11:621-632.

[16] Logue SE, Martin SJ. (2008) Caspase activation cascades in apoptosis. Biochem. Soc. Trans., 36(Pt 1):1–9.

[17] Nakagawa T, Zhu H, Morishima N, Li E, Xu J, Yankner B A, Yuan J. (2000) Caspase-12 mediates endoplasmic reticulum-specific apoptosis and cytotoxicity by amyloid beta. Nature, 403:98-103.

[18] Jäättella M, Tschopp J. (2003) Caspase-independent cell death in T lymphocytes. Nature immunology, 4:416-422.

[19] Kang MH, Reynolds CP (2009) Bcl-2 Inhibitors:targeting mitochondrial apoptotic pathways in cancer therapy. Clin. Cancer Res.,15(4):1126-1132.

[20] Igney FH, Krammer PH. (2002) Death and anti-death: tumour resistance to apoptosis. Nat. Rev. Cancer, 2:277–288.

[21] Marsden V (2002) Apoptosis initiated by Bcl-2- regulated caspase activation independently of the cytochrome *c*/Apaf-1/caspase-9 apoptosome. Nature, 419:634–637.

[22] Elmore S. (2007) Apoptosis: A Review of programmed cell death. Toxicol. Pathol., 35(4):495–516.

[23] Wallach D, Kovalenco AV, Varfolomeev EE, Boldin MP. (1998) Death-inducing fuctions of ligands of the tumor necrosis factor family: a Sanhedrib verdict. Current Opinion in Immunology, 10:279-288.

[24] Salvensen G S, Dixit, VM. (1997) Caspases: Intracellular signaling by proteolysis. Cell, 91:443-446.

[25] Green DR, Reed JC (1998) Mitochondria and apoptosis. Science, *281*:1309–1312.

[26] Reed JC. (2000) Warner Lambert/Parke Davis award lecture: mechanisms of apoptosis. American Journal of Pathology, 157:1415-1430.

[27] Hengartner M, Ellis RE, Horvitz HR. (1992) Caenorhabditis elegans gene ced-9 protects cells from programmed cell death. Nature, 356:494–499.

[28] Evan GI, Wyllie AH, Gilbert CS, Littlewood TD, Land H, Brooks M, Waters CM, Penn LZ, Hanccock DC. (1992) Induction of apoptosis in fibroblasts by c-myc protein. Cell, 69:119-128.
[29] Thornberry N, Lazebnik Y. (1998) Caspases: enemies within. Science, *281*, 1312–1316.
[30] Nicholson W. (1999) Caspase structure, proteolytic substrates, and function during apoptotic cell death. Cell Death Differ., 6:1028–1042.
[31] Martinvalet D, Zhu P, Lieberman J (2005) Granzyme A induces caspase- independent mitochondrial damage, a required first step for apoptosis. Immunity, 22:355–70.
[32] Lassus P, Opitz-Araya X, Lazebnik Y. (2002) Requirement for caspase-2 in stress-induced apoptosis before mitochondrial permeabilization. Science, 297:1352–1354.
[33] Pistritto G, Jost M, Srinivasula SM, Baffa R, Poyet JL, Kari C, Lazebnik Y, Rodeck U, Alnemri ES. (2002) Expression and transcriptional regulation of caspase-14 in simple and complex epithelia. Cell Death Differ. 9:995–1006.
[34] Koenig U, Eckhart L, Tschachler E. (2001) Evidence that Caspase-13 is not a human but a bovine gene. Biochem. Biophys. Res. Commun., 285:1150–1154.
[35] Wajant H (2002) The Fas Signaling Pathway: More than a paradigm. Science, 296(5573):1635-1636.
[36] Wajant H (2003) Death receptors. *Essays* Biochem., 39:53-71.
[37] Gupta S (1996) Apoptosis/Programmed Cell Death: A historical Perspective. *In*: P. Press (Ed.) Mechanisms of lymphocyte Activation and Immune Regulation VI, Vol. 406. Plenum Publishing Corporation, New York, p. 270.
[38] Csiszár Á. (2006) Structural and functional diversity of adaptor proteins involved in tyrosine kinase signalling. BioEssays, 28(5):465–479.
[39] Ulukaya E, Acilan C, Yilmaz Y. (2011) Apoptosis: why and how does it occur in biology? Cell Biochem. Funct., 29:468–480.
[40] Peter ME, Krammer PH. (1998) Mechanisms of CD95 (APO-1/Fas) mediated apoptosis. Current Opinion Immunology, 10:545–551.
[41] Green DR, Droin N, Pinkoski M. *(2003)* Activation-induced cell death in T cells. Immunological Rev., 193:70–81.
[42] Degterev A, Boyce M, Yuan J. (2003) A decade of caspases. Oncogene, 22, 8543–8567.
[43] Degli-Esposti, M. (1999) To die or not to die-the quest of the TRAIL receptors. J. Leukoc. Biol., 65:535–542.
[44] Pan G, O'Rourke K, Chinnaiyan AM. (1997) The receptor for the cytotoxic ligand TRAIL. Science, 276:111–113.
[45] Ashkenazi, A., Dixit VM. (1998) Death receptors: signaling and modulation. Science, 281:1305–1308.
[46] Suliman A, Lam A, Datta R, Srivastava RK. (2001) Intracellular mechanisms of TRAIL: apoptosis through mitochondrial-dependent and -independent pathways. Oncogene, 20:2122–2133.
[47] Budihardjo I, Oliver H, Lutter M, Luo X, Wang X. (1999) Biochemical pathways of caspase activation during apoptosis. Annu. Rev. Cell Dev. Biol., *15*, 269–290.
[48] Hunot S, Flavell RA. (2001) Apoptosis. Death of a monopoly? Science, 292: 865–866.
[49] Von Ahsen O, Renken C, Perkins G, Kluck RM, Bossy-Wetzel E, Newmeyer DD (2000) Preservation of mitochondrial structure and function after Bid- or Bax mediated cytochrome c release. J Cell Biol 150:1027-1036.

[50] Chipuk JE, Kuwana T, Bouchier-Hayes L. (2004) Direct activation of Bax by p53 mediates mitochondrial membrane permeabilization and apoptosis. Science, 303:1010-1014.
[51] Daniel PT, Krammer PH. (1994) Activation induces sensitivity toward APO-1 (CD95)-mediated apoptosis in human B cells. J. Immunol., 152:5624–5632.
[52] Opferman JT, Korsmeyer SJ. (2003) Apoptosis in the development and maintenance of the immune system. Nature immunology, 4(5):410-415.
[53] Hengartner MO. (2000) The biochemistry of apoptosis. Nature, 407(6805):770–776.
[54] Finkel E. (2001) The mitochondrion: Is it central to apoptosis. Science, 292(5517):624–626.
[55] Srinivasula SM, Hegde R, Saleh A. (2001) A conserved XIAP-interaction motif in caspase-9 and Smac/DIABLO regulates caspase activity and apoptosis. Nature, 410:112–116.
[56] Krammer PH (2000) CD95's deadly mission in the immune system. Nature, 407:789–795.
[57] Adams JM. (2003) Ways of dying: multiple pathways to apoptosis. Genes Dev., 17:2481–24.
[58] Czerski L, Nunes G. (2004) Apoptosome formation and Caspase activation: is it different in the heart. J. Mol. Cell Cardiol., 37(3):643-652.
[59] O'Brien Mauria A, Kirby R. (2008) Apoptosis: A review of pro-apoptotic and anti-apoptotic pathways and dysregulation in disease. J. Vet. Emerg. Crit. Care, 18(6):572–585.
[60] Zimmermann KC, Green DR. (2001) How cells die: apoptosis pathways. J. Allergy Clin. Immunol., 108(Suppl 4):S99-S103.
[61] Fesik SW, Shi Y. (2001) Controlling the Caspases. Science, 294(5546):1477-1478.
[62] Kang MH, Reynolds CP. (2009) Bcl-2 Inhibitors:Targeting Mitochondrial Apoptotic Pathways in Cancer Therapy Clin. Cancer Res.,15(4):1126-1132.
[63] Packham G, Stevenson FK (2005) Bodyguards and assassins: Bcl-2 family proteins and apoptosis control in chronic lymphocytic leukemia. Immunology, 114:441-449.
[64] Youle, R J. Strasser A. (2008) The BCL-2 protein family: opposing activities that mediate cell death. Molecular Cell Biology, 9:47- 59.
[65] Power C, Fanning N, Redmond (2002) Cellular apoptosis and organ injury in sepsis: a review. Shock 18(3):197–211.
[66] Hakem R, *et al* (1998) Differential requirement for caspase 9 in apoptotic pathways *in vivo*. Cell, 94:339–352.
[67] Gupta S. (*2005)* Molecular mechanisms of apoptosis in the cells of the immune system in human aging. Immunol. Rev., 205:114–129.
[68] Zakeri Z, Lockshin RA. (2008) Cell death: history and future. Adv. Exp. Med. Biol., 615:11.
[69] Haupt S, Berger M, Goldberg Z, Haupt Y. (2003) Apoptosis the p53 network. J. Cell Sci., 116(Pt20):4077-4085.
[70] Jameson SC, Hogquist KA, Bevan MJ. (1995) Positive selection of thymocytes. Annu. Rev. Immunol., 13:93–126.
[71] Ashton-Rickardt PG, Tonegawa S. (1994) A differential-avidity model for T-cell selection. *Immunol.Today,* 15:362–366.
[72] Von Boehmer H. (1992) Thymic Selection: A Matter of Life and Death. Immunol. *Today,* 13:454-458.
[73] Owen JJ, Jenkinson EJ (1992) Apoptosis and T-cell repertoire selection in the thymus. Ann. NY Acad. Sci., 663:305–310.
[74] Harris SL, Levine AJ. (2005) The p53 pathway: positive and negative feedback loops. Oncogene, 24: 2899–2908.

[75] Krammer PH, Behrmann I, Daniel P, Dhein J, Debatin KM. (1994) Regulation of apoptosis in the immune system. Current Opinion in Immunology, 6:279-289.
[76] Scaffidi C, Kirchhoff S, Krammer PH, Peter ME. (1999) Apoptosis signaling in lymphocytes. Current Opinion immunology, 11:277-285.
[77] Sebzda E, Mariathasan S, Ohteki T, Jones R, Bachmann MF, Ohashi O (1999) Selection of the T Cell Repertoire. Annual Review of Immunology 17: 829-874.
[78] Berzins SP, Godfrey DI, Miller JFAP, Boyd RL. (1999) A central role for thymic emigrants in peripheral T cell homeostasis. Proc. Natl. Acad. Sci. USA, 96:9787–9791.
[79] Krammer, P. H. (1999) CD95(APO-1/Fas)-mediated apoptosis: live and let die. Adv. Immunol., 71:163-210.
[80] Xu G, Yufang S. (2007) Apoptosis signaling pathways and lymphocyte homeostasis. Cell Research 17(9):759-771.
[81] Zhang N, Hartig H, Dzhagalov I, Draper D, HE YW (2005) The role of apoptosis in the development and function of T lymphocytes. Cell Research, 15(10):749-769
[82] Egerton M, Scollary R, Shortman K. (1990) Kinetic of mature T cell development in the thymus. Proc. Natl. Acad. Sci.USA, 87:2579-2582.
[83] Surth CD, Sprent J. (1994) In situ detection of T-cell apoptosis during positive and negative selection in the thymus. Nature, 372:100-103.
[84] Norbury CJ, Hickson ID. (2001) Cellular responses to DNA damage. Annu. Rev. Pharmacol. Toxicol., 41:367–401.
[85] Strasser A, Pellegrini M. (2004) T-lymphocyte death during shutdown of an immune response. Trends in Immunology, 25(11):610-615.
[86] Osmond DG, Kim N, Manoukian R, Phillips RA, Rico-Vargas SA, Jacobsen K. (1992) Dynamics and localization of early B-lymphocyte precursor cells (pro-B cells) in the bone marrow of scid mice. Blood, 79:1695-1703
[87] Krammer PH, Dhein J, Walczak H, Behrmann I, Mariani S, Matiba B, Fath M, Daniel PT, Knipping E, Westendorp MO, Stricker K, Bäumler C, Hellbardt S, Germer M, Peter ME and, Debati KM. (1994) The Role of APO-1-Mediated Apoptosis in the Immune System. Immunological Reviews, 142(1):175-191.
[88] Singer GG, Abbas AK. (1994) The Fas antigen is involved in peripheral but not thymic deletion of T lymphocytes in T cell receptor transgenic mice. Immunity, 1(5): 365–371.
[89] Kabelitz D, Pohl T, Pechhold K. (1993) Activation-induced cell death (apoptosis) of mature peripheral T lymphocytes. Immunol. Today, 14:338–339.
[90] Green D R, Scott DW. (1994) Activation-induced apoptosis in lymphocytes. Current Opinion in Immunology, 6:476-487.
[91] Jones LA, Chin LT, Longo DL, Kruisbeek AM. (1990) Peripheral clonal elimination of functional T cells. Science. 250:1726-1729.
[92] Newell MK, Haughn LJ, Maroun CR, Julius MH. (1990) Death of mature cells by separate ligation of CD4 and the T-cell receptor for antigen. Nature, 347:286-289.
[93] Wesselborg S, Janssen O, Kabelitz D. (1993) Induction of activation-driven death (apoptosis) in activated but not resting peripheral blood T cells. J. Immunology, 150:4338-4345.
[94] Thome M, Schneider P, Hofman NK, Fickenscher H, Meinl E, Neipel F, Mattmann C, Burns K, Bodmer JL, Schroter M, Scaffifidi C, Krammer PH, Peter ME, Tschopps J.

(1997) Viral FLICE-inhibitory proteins (FLIPs) prevent apoptosis induced by death receptors. Nature, 386:517-521.

[95] Janssen O, Qian J, Linkermann A, Kabelitz D. (2003) CD95 ligand – death factor costimulatory molecule? Cell Death Differ. 10:1215–1225.

[96] Janssen O, Sanzenbacher R, Kabelitz D. (2000) Regulation of activation-induced cell death of mature T-lymphocyte populations. Cell Tissue Res., 301:85–99.

[97] Hildeman DA, Zhu Y, Mitchel TC, Kappler J, Marrack P. (2002) Molecular mechanisms of activated T cell death in vivo. Current Opinion in Immunology, 14:354-359.

[98] Zhang N, Hartig H, Dzhagalov I, Draper D, Wen HE. (2005) The role of apoptosis in the development and function of T lymphocytes. Cell Res., 15(10):749-769.

[99] Ruiz-Ruiz MC, Oliver FJ, Izquierdo M, López-Rivas A. (1995) Activation-induced apoptosis in Jurkat cells through a myc-independent mechanism. Molecular Immunology, 32(13):947–955.

[100] Alam A, Cohen LY, Aouad S, Sekaly RP. (1999) Early activation of caspases during T lymphocyte stimulation results in selective substrate cleavage in nonapoptotic cells. J. Exp. Medical. 190:1879-1890.

[101] Yonehara S, Ishii A, Yonehara M. (1989) A cell-killing monoclonal antibody (anti-Fas) to a cell surface antigen co-downregulated with the receptor for tumor necrosis factor. J. Exp. Med., 169:1747–1756.

[102] Trauth BC, Klas C, Peters AM, Matzku S, Möller P, Falk W, Debatin KM, Krammer PH. (1989) Monoclonal antibody-mediated tumor regression by induction of apoptosis. Science, 245:301–305.

[103] Bajorath J, Aruffo A (1997) Prediction of the three-dimensional structure of the human Fas receptor by comparative molecular modeling. J. Comput. Aided. Mol. Des., 11:3–8.

[104] Callard R, Hodgkin P (2007) Modeling T- and B-cell growth and differentiation. Immunological Reviews, 216:119-129.

[105] Malissen B, Malissen M. (1996) Functions of TCR and pre-TCR subunits: lessons from gene ablation. Current Opinion Immunol., 8:383-393.

[106] Samelson LE, Phillips AF, Luong ET, Klausner RD. (1990) Association of the fyn protein-tyrosine kinase with the T-cell antigen receptor. Proc. Natl. Acad. Sci. USA, 87: 4358–4362.

[107] Klausner RD, Patel MD, O'Shea JJ, Samelson LE. (1987) Phosphorylation of the T cell antigen receptor: multiple signal transduction pathways. J. Cell Physiol. Suppl., 5:49-51.

[108] Thorburn A. (2004) Death receptor-induced cell killing. Cell Signal, 16, 139-144.

[109] Alberola-Ila J, Takaki S, Kerner JD, Perlmutter RM (1997) Differential signaling by lymphocyte antigen receptors. Annual Rev. Immunol., 15:125-154.

[110] O'Reilly LA, Tai L, Lee L, Kruse EA, Grabow S, Fairlie WD, Haynes NM, Tarlinton DM, Zhang JG, Belz GT, Smyth MJ, Bouillet P, Robb L, Strasser, A. (2009) Membrane-bound Fas ligand only is essential for Fas-induced apoptosis. Nature, 461:659–663.

[111] Dhein J, Walczak, Baumler C, Debatin KM, Krammer, PH. (1995) Autocrine T-cell suicide mediated by APO-1/(Fas/CD95). Nature, 373:438–441.

[112] Paulsen M, Valentin S, Mathew B, Adam-Klages S, Bertsch U, Lavrik I, Krammer PH, Kabelitz D, Janssen O. (2010) Modulation of CD4+ T cell activation by CD95 costimulation. Cell Death Differ., doi:10.1038/cdd.2010.134

[113] Lettau M, Paulsen M, Schmidt H, Janssen O. (2011) Insights into the molecular regulation of FasL (CD178) biology. Eur J. Cell Biol., 90:456–466.
[114] Tourneur L, Chiocchia G. (2010) FADD: a regulator of life and death. Trends in Immunology, 31(7):260-269.
[115] Krammer *PH,* Arnold *R,* Lavrik IN. (2007) Life and death in peripheral T cells. Nature Rev. Immunol 7: 532-542.
[116] Okada T, Cyster JG. (2006) B cell migration and interactions in the early phase of antibody responses. Current Opinion in Immunology 18:1–8.
[117] Tarlinton DM, and. Smith KGC. (2000) Dissecting affinity maturation: a model explaining selection of antibody-forming cells and memory B cells in the germinal center. Immunol. Today, 21:436.
[118] Lindhout E, Koopman G, Pals ST, Groot de C (1997) Triple check for antigen specificity of B cells during germinal center reactions. Immunol. Today, 18:573.
[119] Sixt M, Kanazawa N, Selg M, Samson T, Roos G, Reinhardt DP, Pabst R, Lutz MB, Sorokin L. (2005) The conduit system transports soluble antigens from the afferent lymph to resident dendritic cells in the T cell area of the lymph node. Immunity, 22:19–29.
[120] Vikstrom I, Tarlinton DM (2011) B cell memory and the role of apoptosis in its formation. Molecular Immunology 48:1301–1306.
[121] MacLennan IC, Toellner KM, Cunningham AF, Serre K, Sze DM, Zuniga E, Cook MC, Vinuesa CG (2003) Extrafollicular antibody responses. Immunol. Rev.,194:8–18.
[122] Liu YJ, Joshua DE, Williams GT, Smith CA, Gordon J, Maclennan IC. (1989) Mechanism of antigen-driven selection in germinal centers. Nature, 342:929-931.
[123] Batista FD, Harwood NE. (2009) The who, how and where of antigen presentation to B cells. Nature Reviews Immunology 9:15-27.
[124] Manceney M, Richard Y, Colaud D, Tursz T, Wiew J (1991) CD77: an antigen of germinal center B cells entering apoptosis. J. Immunol, 21:1131-1140.
[125] Liu YJ, Cairns JA, Holder MJ, Ahrot SD, Jansen KU, Bonnefoy JY, Gordon J, Maclennan IC. (1991) Recombinant 25-kDa CD23 and Interleukin 1 promote the survival of germinal center B: Evidence for bifurcation in the development of centrocytes rescued from apoptosis. Eur. J. Immunol, 21:1107-1114.
[126] Kuklina E M, Shirshev SV. *(2001)* Role of transcription factor NFAT in the immune response. Biochemistry (Moscow), *66(5): 467-475.*
[127] Hehlgans T, Pfeffer K. (2005) The intriguing biology of the tumour necrosis factor/tumour necrosis factor receptor superfamily: players, rules and the games. Immunology. 115(1):1-20.
[128] Yang E, Zha J, Jockel J, Korsmeyer SJ. (1995) Bad: a heterodimeric partner of Bcl-xL and Bcl-2, displaces Bax and promotes cell death. Cell, 80:285-291
[129] Kondo M, Akashi K, Domen J, Sugamura K, Weissman IL. (1997) Bcl-2 Rescues T lymphopoiesis, but not B or NK cell development, in common γ Chain–deficient mice. Immunity, 7 (1):155–162.
[130] Strasser A, Harris AW, Huang DCS, Krammer PH, Cory S. (1995) Bcl-2 and Fas/APO-1 regulate distinct pathways to lymphocyte apoptosis. EMBO J, 14:6136–6147.
[131] Kearney ER, Pape KA, Lo DY, Jenkins MK (1994) Visualization of peptide-specific T cell immunity and peripheral tolerance induction in vivo. Immunity, 1(4): 243-339.

[132] Maclennan IC, Liij YJ, John GD. (1992) Maturation and dispersal of B-Cell clones during T cell-dependent antibody responses. Immunol. Rev., 126:143-161.

[133] Strasser A. (1995) Life and death during lymphocyte development and function: evidence for two distinct killing mechanisms. Current Opinion Immunol., 7:228-234.

[134] Lassus P, Opitz-Araya X and Lazebnik Y. (2002) Requirement for caspase-2 in stress-induced apoptosis before mitochondrial permeabilization. Science, 297:1352–1354.

[135] MacLennan IC; Liu YJ; Oldfield S; Zhang J; Lane PJ. (1990) The evolution of B-cell clones. Current topics in Microbiology and Immunology,159:37-63.

[136] Nutt SL, Tarlinton DM. (2011) Germinal center B and follicular helper T cells: siblings, cousins or just good friends? Nature Immunology, 12:472–477.

[137] Chen A, Zheng G, Tykocinski ML. (2003) Quantitative interplay between activating and pro-apoptotic signals dictates T cell responses. Cellular Immunology, 221:128-137.

[138] Lenardo M, Chan KM, Hornung F, *et al.* (1999) Mature T lymphocyte apoptosis—immune regulation in a dynamic and unpredictable antigenic environment. Annu. Rev. Immunol., 17:221-253.

[139] Van Parijs L, Refaeli Y, Lord J, Nelson BH, Abbas AK, Baltimore D. (1999) Uncoupling IL-2 signals that regulate T cell proliferation, survival, and Fas mediated activation-induced cell death. Immunity, 11:281–288.

[140] Schmitz I, Krueger A, Baumann S, Schulze-Bergkamen H, Krammer PH, Kirchhoff S. (2003) An IL-2-dependent switch between CD95 signaling pathways sensitizes primary human T cells toward CD95-mediated activation-induced cell death. J. Immunol., 171(6):2930-2936.

[141] Refaeli Y, Van Parijs L, London CA, Tschopp J, Abbas AK. (1998) Biochemical mechanisms of IL-2- regulated Fas-mediated T cell apoptosis. Immunity, 8: 615–623.

[142] Sytwu HK, Liblau RS, McDevitt HO. (1996) The roles of Fas/APO-1 (CD95) and TNF in antigen-induced programmed cell death in T cell receptor transgenic mice. Immunity, 5:17–30.

[143] Janssen EM, Droin NM, Lemmens EE, Pinkoski MJ, Bensinger SJ, Ehst BD, Griffith TS, Green DR, Schoenberger SP (2005) CD4+ T-cell help controls CD8+ T-cell memory via TRAIL-mediated activation-induced cell death. Nature, 434:88–93.

[144] Martínez-Lorenzo MJ, Alava MA, Gamen S, Kim KJ, Chuntharapai A, Pineiro A. (1998) Involvement of APO2 ligand TRAIL in activation-induced death of Jurkat and human peripheral blood T cells. *Eur. J. Immunol.* 28:2714–2725.

[145] Devadas S, Das J, Liu C, Zhang L, Roberts, Pan Z, Moore PA, Gobardhan D. (2006) Granzyme B is critical for T cell receptor-induced cell death of type 2 helper T cells. *Immunity* 25:237–247.

[146] Lenardo MJ. (1991) Interleukin-2 programs mouse αβ T lymphocytes for apoptosis. *Nature* 353, 858–861.

[147] Patarroyo JH, Vargas MI , González CZ , Guzmán F , Martins-Filho OA, Afonso LCC, Valente FL, Peconick AP, Marciano AP, Patarroyo AM V, Sossai S. (2009) Immune response of bovines stimulated by synthetic vaccine SBm7462 against Rhipicephalus (Boophilus) microplus. Vet. Parasitol. 166:333-339.

Chapter 3

Neuronal Apoptosis in HIV-1-Associated Central Nervous Diseases and Neuropathic Pain

Mona Desai, Ningjie Hu, Daniel Byrd and Qigui Yu

Additional information is available at the end of the chapter

1. Introduction

The human immunodeficiency virus type 1 (HIV-1) pandemic has claimed over 20 million lives, with 38.6 million people worldwide currently infected (2009 AIDS Epidemic Update by UNAIDS/WHO, www.unaids.org), and will continue to contribute to human morbidity and mortality. The number of people living with HIV-1 continues to rise worldwide because of high rates of new infections and the success of antiretroviral therapy (ART, also known as the highly active antiretroviral therapy or HAART). HIV-1 infection results in a variety of syndromes involving both the central and peripheral nervous systems. HIV-1 invades the central nervous system (CNS) early during the course of infection and persists thereafter in the absence of treatment. HIV-1 infection of the CNS is often associated with neurological complications including HIV-1-associated severe and mild neurologic disorders. Prior to HAART, neurologic disorders were the first manifestation of symptomatic HIV-1 infection, affecting roughly 10% – 20% of patients and up to 60% of patients in the advanced stages of acquired immunodeficiency syndrome (AIDS)[1]. HIV-1 infection is also associated with neuropathic pain, which is caused by peripheral nervous system injury. The vast majority (up to 90%) of individuals infected with HIV-1 have pain syndromes that significantly impact their well-being[2-13]. A variety of pain syndromes including peripheral neuropathies, headache, oral and pharyngeal pain, abdominal and chest pain, arthralgias and myalgias as well as pain related to HIV-1/AIDS-associated malignancies such as Kaposi's sarcoma[14,15]. Pain occurs at all stages of HIV-1 infection, although its severity and frequency are correlated with disease progression[2,16]. In the era of HAART, patients infected with HIV-1 live longer, and management of their symptoms including pain has emerged as a top priority for HIV-1 clinical and translational research.

Uniquely, neuronal injury, cell loss and dysfunction during HIV-1 infection appear to occur through soluble neurotoxins rather than productive virus infection, because there is no evidence that HIV-1 directly infects neurons. Apoptosis, an active process of cell death characterized by cell shrinkage, chromatin aggregation with genomic fragmentation and nuclear pyknosis, appears to be an important feature of HIV-1-associated central neurological dysfunction and peripheral neuropathy. Apoptotic neurons have been observed in the CNS of HIV-1-infected individuals, and are more abundant in HIV-1 patients with peripheral neuropathy. Herein, we review the role of neuronal apoptosis in HIV-1-associated neurological disorders, focusing on HIV-1-associated CNS manifestations and sensory neuropathic pain. This review also summarizes the current data supporting both the direct and indirect mechanisms by which neuronal apoptosis may occur during HIV-1 infection. Finally, we discuss recent and past approaches for the prevention and treatment of HIV-1-associated neurological disorders by targeting specific neurotoxic signaling pathways.

2. Overview of HIV-1-associated neurological disorders

Neurologic disorders are among the most frequent and devastating complications of HIV-1 infection. Prior to HAART, neurologic disorders were the first manifestation of symptomatic HIV-1 infection, affecting roughly 10% – 20% of patients and up to 60% of patients in the advanced stages of HIV-1/AIDS[1]. The incidence of subclinical neurologic disease is much higher. Autopsy studies of AIDS patients have demonstrated pathologic abnormalities of the nervous system in 75 - 90% of cases[17,18]. Neurologic disorders continue to be prevalent in the era of HAART. HAART has reduced the incidence of severe forms of AIDS-associated neurologic disorders such as HIV-1-associated dementia (HAD), but with longer life span, the prevalence of milder forms of neurologic manifestations such as HIV-1-associated neurocognitive disorder (HAND) appears to be increasing[19,20]. In resource-rich settings such as the United States and the European Union, where antiretroviral therapy is relatively available, peripheral neuropathy and HIV-1-associated cognitive dysfunction (including HAD) account for the greatest proportion of neurologic disease burden[21,22]. In resource-poor settings such as developing countries, opportunistic infections of CNS account for most of the reported neurologic morbidity and mortality in AIDS patients[23]. Fulminant bacterial meningitis, cryptococcal meningitis, neurotuberculosis, neurosyphilis, and toxoplasmosis are common among HIV-1-infected individuals in Asia and Africa[23].

The neuropathic syndromes associated with HIV-1 infection are diverse and include both the somatic and autonomic nervous systems. The primary pathological abnormality may be demyelination, both acute and chronic, or axonal degeneration. Either single or multiple nerves may be involved leading to mono- or polyneuropathy. HIV-1 invades the CNS early in the infectious course, and eventually causes mild or severe forms of HIV-1-associated neurologic manifestations[24]. Severe manifestations include HIV-1-assocated dementia complex (HAD, HIV-1-encephalopathy, and subacute encephalitis) and HIV-1-associated myelopathy, whereas mild manifestations include HIV-1-associated neurocognitive/motor disorders and HIV-1-associated neurobehavioral abnormalities (Table 1)[25]. HIV-1 also affects the peripheral nervous system (PNS) and PNS damage can be assessed by electroneurographic examinations quantifying nerve conduction velocity (NCV) and action potential amplitudes. Defined disorders of the PNS comprise the HIV-1-associated Guillain-Barré syndrome and sensory polyneuropathy (Table 1)[26].

CNS disorders	PNS disorders
Severe manifestations	HIV-1-associated Guillain-Barré syndrome
HIV-1-assocated dementia complex	HIV-1-associated sensory polyneuropathy
HIV-1-associated dementia (HAD)	
Subacute encephalitis	
HIV-1-encephalopathy	
HIV-1-assodated encephalopathy (HIVE)	
HIV-1-associated myelopathy (HIVM)	
Mild manifestations	
HIV-1-associated neurocognitive/motor disorder	
HIV-1-associated neurobehavioral abnormalities	

Table 1. HIV-1-associated CNS and PNS disorders
HAD: HIV-1-associated dementia; HIVE: HIV-1-associated encephalopathy; HIVM: HIV-1-associated myelopathy.

3. Neuronal apoptosis in the central nervous system in HIV-1 infection

Apoptosis of neurons and non-neuronal cells has been demonstrated in the brain of HAD patients[27]. Petito and Roberts reported that neuronal and astrocytic death in HIV-1 infection occurred by apoptosis[27]. They identified apoptotic neurons, astocytes, and multinucleated giant cells in the brain of adults with HIV-1-associated encephalopathy (HIVE) using a combined approach of in situ end labeling (ISEL) and immunohistochemistry[27]. Neuronal apoptosis occurs not only in HIV-1-infected adults but also in infected children. Using an *in situ* technique, Gelbard et al found that newly cleaved 3'-OH ends of DNA, a marker for apoptosis, in the brain of children with HIVE[28]. They demonstrated the presence of apoptotic neurons in cerebral cortex and basal ganglia of children that had HIVE with progressive encephalopathy[28]. In addition, association of the localization of apoptotic neurons with perivascular inflammatory cell infiltrates containing HIV-1 infected macrophages and multinucleated giant cells was observed[28]. Adle-Biassette et al found apoptotic perivascular cells in the brain of adults with HIVE[29]. Neuronal apoptosis was more severe in atrophic brains, and did not directly correlate with productive HIV-1 infection, suggesting that an indirect mechanism of neuronal damage exists. Shi et al demonstrated neuronal apoptosis in brain tissue from AIDS patients and in HIV-1-infected primary human fetal brain cultures[30]. HIV-1 infection of primary brain cultures induced apoptosis in neurons and astrocytes *in vitro* as determined by terminal deoxynucleotidyl transferase-mediated dUTP nick end labeling (TUNEL) and propidium iodide (PI) staining and by electron microscopy[30]. Apoptosis was not significantly induced until 1 - 2 weeks after the time of peak virus production, suggesting that apoptosis of neurons and non-neuronal cells in HIV-1 infection is triggered by soluble factors rather than by direct viral infection[30]. An et al used ISEL technique to examine neuronal apoptosis in the brains of AIDS and pre-

AIDS patients[31,32]. The presence of apoptotic cells in the brains of HIV-1-positive pre-AIDS individuals was observed, although the frequency was lower than that in the brain of AIDS patients. These data suggest that brain damage already occurs during the early stages of HIV-1 infection[31,32].

Apoptotic neurons are found in several regions of the brain of adults with HIVE including the frontal and temporal cortex, basal ganglia and brain stem[27-32]. The basal ganglia apoptotic neurons were found to be more abundant in the vicinity of activated microglia that had higher HIV-1 copies as determined by measuring HIV-1 core proteins[31,32]. Within the cerebral cortex, the extent of neuronal apoptosis correlated with cerebral atrophy patients[31,32]. In addition, apoptotic neurons in the basal ganglia and cerebral cortex of children with HIVE were detected in the vicinity of perivascular inflammatory cell infiltrates containing HIV-1- infected multinucleated giant cells and macrophages[26].

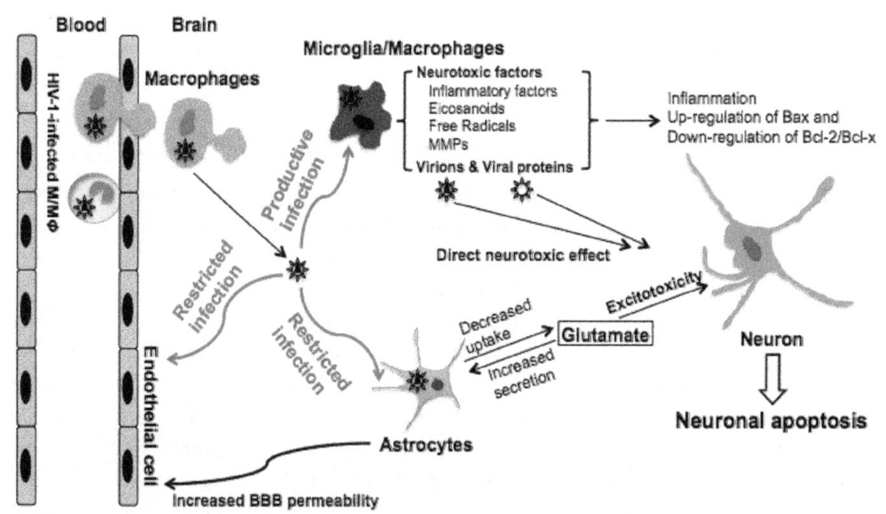

Note: This figure was modified from Jones G. & Power C. Neurobiology of Disease, 2006; 1 – 17
M/MΦ: monocytes/macrophages

Figure 1. Neuronal apoptosis in the central nervous system in HIV-1 infection. Monocytes/macrophages (M/MΦ) play a crucial role in HIV-1-associated neurologic disorders. They are among the first cells infected by HIV-1, which then forms a reservoir of HIV-1 in infected individuals. HIV-1-infected monocytes/macrophages serve as a means (Trojan horse) of spreading the virus to other tissues such as the brain. HIV-1 gains entry into the brain within macrophages that efficiently produce new virions. These virions infect microglia, astrocytes, endothelial cells, and multinucleated giant cells in the brain. Virions and viral proteins, and the neurotoxic factors released from HIV-1-infected cells induce apoptosis of neurons. The disruption of glutamate homeostasis induces neuronal excitotoxicity, resulting in neuronal cell death.

In addition to apoptotic neurons, apoptosis of other CNS cell types in the brain of HAD patients has also been reported. Apoptotic astrocytes were detected and found to be more common in the brain of HAD patients compared to non-demented HIV-1/AIDS patients[27-32]. In addition, greater numbers of apoptotic astrocytes were detected in the brains of AIDS patients with rapidly progressing dementia compared to slow progressors[28,29]. Astrocytes are the most abundant cells of the human brain, and perform many functions including biophysical support of endothelial cells that form the blood–brain barrier (BBB), provision of nutrients to the nervous tissue, maintenance of extracellular ion balance, and a role in the repair and scarring process of the brain and spinal cord following traumatic injuries. HIV-1-mediated loss of astrocytes impairs maintenance of the BBB and alters the composition of the extracellular environment, resulting in increased BBB permeability. These findings suggest that astrocyte cell loss plays a critical role in the neuropathogenesis of HAD (Figure 1). Table 2 summarizes the studies *in vivo* and *in vitro* on neuronal apoptosis in HIV-1 infection.

***In vivo* studies**	
Petito *et al.*, 1995	Apoptotic neurons, astrocytes, and multinucleated giant cells in the brain of adults with HIVE
Gelbard *et al.*, 1995,	Apoptotic neurons, macrophages, and microglia in the brain of children with HIVE
Adie-Biassette *et al.*, 1995	Apoptotic neurons and perivascular cells in the brain of adults with HIVE
Shi *et al.*,1996	Apoptotic neurons, astrocytes, and endothelial cells in the brain of adults with HIVE
An *et al.*, 1996	Apoptotic neurons and glial cells in the brain of AIDS patients and pre-AIDS patients
Krajewski *et al.*, 1997	Elevated numbers of Bax-positive microglia and macrophages in the brain of children with HIVE
Vallat *et al.*, 1998	Apoptotic neurons, astrocytes, endothelial cells, pericytes, and macrophages in the brain of AIDS patients
***In vitro* studies**	
Shi *et al.*, 1996	Apoptosis of neurons and astrocytes in primary human brain cultures infected with HIV-$1_{89.6}$
New *et al.*, 1997	Apoptosis of primary human neurons induced by soluble HIV-1 Tat protein
Talley *et al.*, 1995	Apoptosis of differentiated SK-N-MC human neuroblastoma cells induced by TNF-α
Muller *et al.*, 1992	Apoptosis of neurons in rat cortical cultures induced by soluble HIV-1 gp120 protein
Chi *et al.*, 2011	Apoptosis of rat primary dorsal root ganglion neurons directly induced by HIV-1 Tat protein

This table was modified from Shi et al., the Journal of NeuroVirology, 1998: 4, 281 - 290

Table 2. Summary of *in vitro* and *in vivo* studies on neuronal apaoptsois in HIV-1 infection

4. Neuronal apoptosis is associated with pain in HIV-1-infected patients.

The vast majority (up to 90%) of individuals living with HIV-1 have pain syndromes that significantly impact their well-being[2-13]. Pain occurs at all stages of HIV-1 infection, although its severity and frequency are correlated with disease progression[2,16]. In the era of HAART, patients infected with HIV-1 live longer, and management of their symptoms including pain is emerging as a top priority for HIV-1 clinical and translational research. However, pain is often under-assessed and undertreated in people with HIV-1/AIDS, and little progress has been made in understanding the underlying mechanisms that are key for pain management and improvement of quality of life.

The pain in HIV-1-infected patients is believed to result from (1) direct neurotoxic effects of viral components on neurons in both central and peripheral nervous systems, (2) immune dysregulation leading to inflammatory changes, opportunistic infections and/or tumors, and (3) the adverse effects related to antiviral drugs[2,3,33]. Viral proteins are directly involved in neuronal damage and death. HIV-1 encodes a total of nine viral proteins including three structural proteins (Env, Pol, and Gag), two essential regulatory proteins (Tat and Rev), and four accessory proteins (Vif, Vpr, Vpu, and Nef). These viral proteins have been extensively studied for their role in HIV-1-associated CNS neuropathy. To date, five of these proteins including Env, Tat, Vpr, Nef, and Rev have been identified as potent neurotoxic viral proteins in that they directly induce neuronal cell death (Figure 2). These proteins are implicated in HIV-1-associated CNS pathologies such as mild to severe cognitive impairments and encephalitis[34-43]. Recent studies have shown that Env and Vpr exert neurotoxic activities on peripheral sensory neurons (pain-sensing neurons). Injection of Env into the spinal intrathecal space of rats causes marked pain-like effects[44-48], and Vpr enhances excitability of dorsal root ganglion (DRG) neurons[49]. These results suggest that HIV-1 neurotoxic proteins have direct effects on peripheral nerves, and may be causative factors in the generation of neuropathic pain in HIV-1-infected individuals.

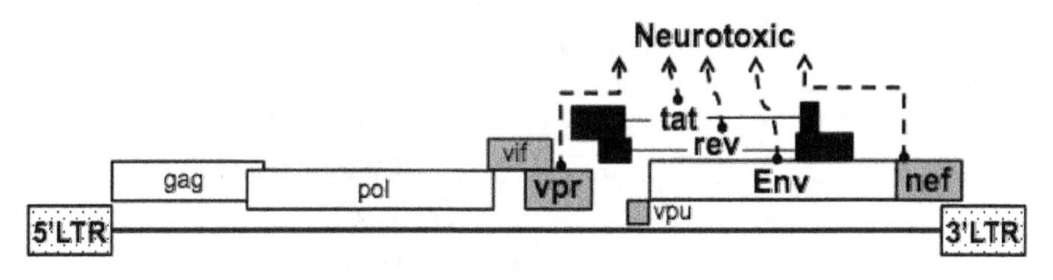

Figure 2. HIV-1 genome and viral proteins. HIV-1 is composed of two copies of single-stranded RNA genome with approximately 9700 bp in length. HIV-1 genome encodes a total of nine viral proteins including three structural proteins (Env, Pol, and Gag in open boxes), two essential regulatory proteins (Tat and Rev in black boxes), and four accessory proteins (Vif, Vpr, Vpu, and Nef in gray boxes). Genes encoding HIV-1 proteins with know neurotoxicity are given in bold.

HIV-1 Env gene encodes a 160-kDa envelope glycoprotein (gp160) precursor, which is proteolytically cleaved into the exterior (gp120) and transmembrane (gp41) glycoproteins. The gp120 remains associated with the mature envelope glycoprotein complex through a non-covalent interaction with the gp41 ectodomain. Three exterior gp120 glycoproteins and three transmembrane gp41 proteins are assembled as a trimer by non-covalent interactions. The gp120 has been shown to directly interact with neurons, leading to neuronal apoptosis. *In vivo* experiments have shown that subcutaneous injection of purified recombinant gp120 in neonatal rats causes dystrophic changes in pyramidal neurons of cerebral cortex accompanied by abnormalities of developmental behaviors[50]. Toggas et al generated transgenic mice expressing gp120 mRNA in astrocytes and found a spectrum of neuronal and glial changes resembling abnormalities in brains of HIV-1-infected individuals[51]. The severity of brain damage correlated positively with the brain level of gp120 expression[51]. These results provide *in vivo* evidence that gp120 plays a key role in HIV-1-associated nervous system impairment. Damage of neurons has also been reported in transgenic mice expressing the entire HIV-1 genome[52], Tat[53], or Vpr[54]. Transgenic mice expressing Nef[55,56] primarily showed severe impairment of thymocyte development and peripheral T-cell function by a CD4-independent mechanism. Nef expression in these mice also indirectly contributes to HIV-1-associated neuropathy by enhancing protein expression of other HIV-1 genes and altering cellular maturation *in vivo*[57]. *In vitro* exposure to HIV-1 proteins including gp160, gp120, gp41, Tat, Nef, Rev, and Vpr has been reported to initiate neuronal damage[58-62].

Dorsal root ganglion (DRG) neurons have central terminals in the spinal cord dorsal horn and peripheral terminals in skin, muscle and other peripheral tissues. These DRG neurons transmit and relay pain-related signals and temperature sensation from peripheral tissues to the spinal cord and brain. Neurotoxins and inflammatory molecules cause hyperexcitability of primary DRG neurons leading to spontaneous or persistent firing. These agents also induce apoptosis of DRG neurons, resulting in permanent neuron damage, death, and nerve lesions. It is well established that peripheral sensitization of primary DRG neurons is the key event in the onset of chronic pain conditions [63]. Peripheral sensitization is defined as the enhancement of a baseline response, such as action potential (AP) firing, after exposure to a defined mediator [63]. For example, exposure to the pro-inflammatory prostaglandin E_2 (PGE_2), increases the number of APs evoked by an excitatory stimulus by about three-fold compared to the control, thus the neuronal output is intensified by PGE_2 [64,65]. This increased firing is believed to be critical in the enhanced perception of pain sensation for both inflammatory and neuropathic conditions [64,65]. Our group has recently explored the direct effects of HIV-1 Tat protein on excitability of rat primary DRG neurons[62]. We demonstrated that HIV-1 Tat triggered a rapid and sustained enhancement of the excitability of small-diameter rat primary DRG neurons, which was accompanied by marked reductions in the rheobase and resting membrane potential (RMP), and an increase in the resistance at threshold (R_{Th}). Such Tat-induced DRG hyperexcitability may be a consequence of the inhibition of cyclin-dependent kinase 5 (Cdk5) activity. Tat rapidly inhibited Cdk5 kinase activity and mRNA production, and roscovitine, a well-known Cdk5 inhibitor, induced a very similar pattern of DRG hyperexcitability. Indeed, pre-application of Tat prevented roscovitine from having additional effects on the RMP and action potentials (APs) of DRGs. However, Tat-mediated actions on the rheobase and R_{Th} were accelerated by roscovitine. These results suggest that Tat-mediated changes in DRG excitability are partly facilitated by Cdk5 inhibition. In addition, Cdk5 is most abundant in DRG neurons and participates in the regulation of pain signaling. We also demonstrated that HIV-1 Tat markedly induced apoptosis of primary DRG neurons after exposure for longer than 48 h. Together, our work indicates that HIV-1 proteins are capable of producing pain signaling through direct actions on excitability and survival of sensory neurons.

However, we still do not know: (1) whether all five HIV-1 neurotoxic proteins induce pain-related signaling, (2) what is their rank order for induction of pain-related signaling, (3) whether these neurotoxic proteins exert synergistic effects on pain-related signaling, and (4) the molecular mechanisms by which HIV-1 neurotoxic proteins exert neuropathogenic effects on sensory neurons. Clarification of these issues will have potentially important implications for developing therapeutic strategies to prevent or treat HIV-1-assciated pain. For example, early initiation of HAART and/or neutralization of HIV-1 protein neurotoxicity may significantly prevent or delay HIV-1-associated pain.

5. Mechanisms of neuronal apoptosis induced by HIV-1 infection

There are two main apoptotic pathways: the extrinsic or death receptor pathway and the intrinsic or mitochondrial pathway[66,67]. There is an additional pathway that involves T-cell mediated cytotoxicity and perforin-granzyme-dependent killing of the cell by inducing apoptosis via either granzyme B or granzyme A[67]. The extrinsic and intrinsic pathways are initiated by caspase-3 activation induced by proteolytic cleavage and results in DNA fragmentation, degradation of cytoskeletal and nuclear proteins, formation of apoptotic bodies, expression of ligands for phagocytic cell receptors and finally uptake by phagocytic cells, whereas the granzyme pathway activates a parallel, caspase-independent cell death pathway via single stranded DNA damage. All of these three pathways may be involved in neuronal apoptosis in HIV-1 infection.

5.1. Extrinsic pathway

Several factors including neurotoxins such as tumor necrosis factor alpha (TNF-α) secreted by HIV-1–infected macrophages and microglia, and soluble forms of the HIV-1 gp120 and Tat proteins have been reported to mediate HIV-1-induced neuronal injury[30]. TNF-α is elevated in the serum, cerebrospinal fluid (CSF), and brain of AIDS patients. The elevated TNF-α levels have been shown to correlate with clinical dementia[68]. The binding of TNF-α to TNF-α receptor-1 (TNFR1) on the surface of neurons activates TNFR1-associated death domain protein (TRADD), which in turn interacts with Fas-associated death domain protein (FADD) to induce apoptosis of neurons[69]. However, TNF-α and TNF-beta (TNF-β) have also been shown in certain circumstances to be neuroprotective. TNF-α and TNF-β protect cultured embryonic rat hippocampal, septal, and cortical neurons against glucose deprivation-induced injury and excitatory amino acid toxicity[70]. TNF-α induces the expression of the chemokine CX3CL1 in astrocytes[71], which in turn protects neurons from HIV-1 gp120-mediated toxicity[72]. TNF-α protective or destructive effects on neuronal survival depend on the timing, duration, and concomitant expression of other elements.

The TNF-related apoptosis-inducing ligand (TRAIL), a member of the TNF superfamily, is also involved in neuronal apoptosis[73]. TRAIL is a type II integral membrane protein and expressed by multiple cell types[73]. TRAIL protein levels were increased in human monocyte-derived macrophages after HIV-1 infection and immune activation[73]. In the brain of HIVE patients, TRAIL-expressing macrophages were found in association with active caspase-3 positive neurons[73]. *In vitro* studies have shown that TRAIL induces a dose-dependent effect on neuronal apoptosis[73].

Five HIV-1 proteins including Env, Tat, Vpr, Nef, and Rev, are potent neurotoxic viral proteins that cause apoptosis of various neuronal populations in the CNS, resulting in pathologies such as cognitive impairment and encephalitis associated with NeuroAIDS (Figure 2). HIV-1 gp120 binds with high affinity to CXCR4 expressed on human neurons[74], resulting in neuronal apoptosis[40,75]. CXCR4 is a seven transmembrane domain G-protein-coupled receptor and activation of this receptor via binding of gp120 triggers neuronal apoptosis. The exact mechanism underlying gp120-CXCR4-mediated neuronal apoptosis remain unclear, but is not related to increases in intracellular calcium (Ca^{2+}) or to induction of glutamate uptake by functional glutamate receptors[75]. Tat appears to exert its neurotoxic activity via binding to the N-methyl D-aspartate (NMDA) receptor and the low-density lipoprotein receptor-related protein (LRP)[76-79], whereas evidence of Vpr, Nef, or Rev binding to surface receptor(s) of neurons is lacking. Binding of HIV-1 Tat to LRP triggers formation of a macromolecular complex involving the LRP, NMDA receptors, postsynaptic density protein-95 (PSD-95), and neuronal nitric oxide synthase (nNOS) at the neuronal plasma membrane. This complex leads to neuronal and astrocyte apoptosis. Blockade of LRP-mediated Tat uptake, NMDA receptor activation, or nNOS significantly reduces neuronal apoptosis, suggesting that formation of this complex is an early step in Tat-mediated neuronal apoptosis. CCL2, an inflammatory chemokine, inhibits formation of the complex, resulting in protection against Tat-mediated neuronal apoptosis[80].

During HIV-1 infection, multiple viral proteins co-exist in circulation and tissues, and synergize with one another to accelerate disease progression. For example, Env and Tat act synergistically to cause cell death in CNS neurons[81]. Vpr and Nef each can induce injury of podocytes, and have a synergistic effect on podocyte injury and the subsequent development of glomerulosclerosis[82]. In addition, HIV-1 proteins have synergistic effects with inflammatory factors[83], methamphetamine[84,85], and ethanol[86] in acceleratory HIV-1-associated neurological disorders. Thus, it is possible that the known HIV-1 neurotoxic proteins have synergistic effects on neuronal apoptosis.

5.2. Intrinsic pathway

The intrinsic (or mitochondrial) pathway of apoptosis is intracellularly initiated in response to various types of intracellular signals including growth factor withdrawal, DNA damage, unfolding stresses in the endoplasmic reticulum (ER) and death receptor stimulation. The intrinsic apoptotic pathway is characterized by permeabilisation of the mitochondria and release of cytochrome C (Cyt C) into the cytoplasm. Once it is released, Cyt C binds to the cytosolic protein known as apoptotic protease activating factor 1 (Apaf-1) to facilitate the formation of the apoptosome, a large multi-protein complex formed in the process of apoptosis[87,88]. Subsequently, the apoptosome binds and activates caspase-9 preproprotein. Activated caspase-9 then activates downstream caspases resulting in the activation of caspase cascade[87,88].

Studies have shown that HIV-1 Vpr directly interacts with the mitochondrial membrane to induce neuronal apoptosis through the intrinsic pathway[54]. Vpr directly bind to the adenine nucleotide translocator (ANT), a component of the mitochondria permeability transition pore located on the inner mitochondrial membrane, to trigger a rapid dissipation of the mitochondrial transmembrane potential and the mitochondrial release of apoptogenic proteins such as Cyt C or apoptosis inducing factor[89,90]. Vpr induces neuronal cell death *in vivo* in the absence of both profound microglia activation and pro-inflammatory gene expression[54]. In addition, Vpr causes neuronal apoptosis independent of the presence and function of the p53 tumor suppressor[54,91], a transcription factor which regulates the expression of several pro-apoptotic proteins such as Bax, Bad, Puma, and Bid[92]. These findings further indicate that Vpr-induced apoptosis is effected via the intrinsic pathway.

In addition to Vpr, HIV-1 gp120 and Tat also trigger neuronal apoptosis via intrinsic pathway[93]. Both gp120 and Tat enhance p53 expression and phosphorylation, and promote BCL-2-associated X protein (Bax) insertion into the mitochondrial membrane. Peripheral blood mononuclear cells (PBMCs) and lymph node syncytia from HIV-1 patients show accumulation of phosphorylated p53[94]. Accumulation of p53 protein in neurons of HAD patients and in CNS tissues from monkeys with simian immunodeficiency virus (SIV) encephalitis (SIVE) has been reported[95,96]. It is well established that p53 can mediate the intrinsic apoptosis pathway[94]. HIV-1 can establish a productive infection in perivascular macrophages and a subset of parenchymal microglia in the brain. These infected cells secrete gp120 and other immunomodulatory factors that lead to activation of astrocytes and microglia. Accumulation of p53 protein also occurs in neurons and non-neuronal CNS cells, which is required in both neurons and microglia for gp120-induced neuronal apoptosis[97].

The Bcl-2 family of proteins plays a key role in the regulation of apoptosis[98,99]. This gene family is comprised of antiapoptotic members including Bcl-2, Bcl-XL, Bcl-w, Mcl-1, Bfl-1, and Bcl-B, and proapoptotic members including Bax, Bak, Bad, Bok, Bik, Bid, Bim, Hrk, Blk, Bnip3, Noxa, Puma, and Bcl-G[100]. Both Tat and gp120 promote Bax insertion into the mitochondrial membrane and subsequent release of Cyt C[93]. Such effect of HIV-1 gp120 and Tat can be blocked by anti-apoptotic proteins BCL-2/BCL-XL[93]. In addition, Tat also inhibits expression of BCL-2 in neurons, resulting in induction of neuronal apoptosis[101]. *In vitro* studies have shown that over-expression of Bcl-2 in neuronally differentiated human SK-N-MC cells protect neurons against Tat-induced apoptosis[101].

Dysregulation of neuronal cell Ca^{2+} homeostasis plays a central role in both HIV-1 gp120 and Tat-induced neuronal apoptosis through the intrinsic pathway[54,102]. Both gp120 and Tat disrupt neuronal Ca^{2+} homeostasis by perturbing Ca^{2+}-regulating systems in the plasma membrane and ER. By altering voltage-dependent Ca^{2+} channels, glutamate receptor channels, and membrane transporters, HIV-1 gp120 and Tat promote Ca^{2+} overload, oxyradical production, and mitochondrial dysfunction. Exposure to gp120 induces an increased release of arachidonic acid in rat primary neuronal cell culture followed by NMDA receptor-mediated neurotoxicity[103]. The elevated arachidonic acid impairs metabolic balance of glutamate in neurons, resulting in lethal levels of Ca^{2+} influx. As evidence, gp120-induced neurotoxicity is inhibited by dizocilpine and memantine, two blockers of NMDA receptors, and by AP5, a competitive antagonist of the glutamate-binding site on NMDA receptors, but not by antagonists of non-NMDA receptors, which are less permeable to calcium[81,102]. Similar to gp120, Tat activates a wide variety of intracellular signals, some of which play a critical role in neuronal apoptosis. Tat causes activation of c-Jun N-terminal kinases (JNKs), activator protein-1 (AP-1), and phosphatidylinositol 3-kinases (PI3Ks) in a time- and dose-dependent manner[104,105], leading to increases in activity of the protein kinase C (PKC) and mitogen-activated protein (MAP) kinase[104]. Intact molecules of Tat and peptide fragments of Tat, in particular, Tat_{31-61}, increased intracellular Ca^{2+} in cultured human fetal neurons[106]. The Ca^{2+} influx contributed to Tat-induced neuron depolarization by activation of glutamate receptors[106]. Tat is also able to increase levels of inositol 1,4,5-triphosphate (IP_3) that in turn triggers Ca^{2+} release from IP_3-sensitive ER stores in cultured human fetal astrocytes and neurons[107]. Inhibition of IP_3-mediated Ca^{2+} release from the ER has been shown to inhibit Tat-induced neurotoxicity[107]. Notably, HIV-1 gp120 and Tat have a synergistic effect on Ca^{2+} dysregulation in neurons[81]. Subtoxic concentrations of gp120 and Tat causes prolonged increases in levels of intracellular Ca^{2+}, resulting in neuronal cell death. The memantine, a potent NMDA receptor blocker, can completely block the neurotoxicity caused by Tat and gp120 applied in combination[81].

Both HIV-1 gp120 and Tat directly stimulate neurons and non-neuronal cells in the brain to produce inflammatory factors including TNF-α, interleukin (IL)-6 (IL-6), IL-8, and CXCL10 (also known as interferon gamma-induced protein 10 or IP-10)[47,108-110]. These inflammatory factors cause excessive Ca^{2+} influx and oxidative stress, resulting in neuronal apoptosis.

Notably, there is crosstalk between the extrinsic and intrinsic pathways[67]. For example, cleavage of the BCL-2-family member Bid by caspase-8 activates the mitochondrial pathway after apoptosis induction through death receptors, and can be used to amplify the apoptotic signal[111].

6. Future perspectives: Implications for therapy

As discussed, productive infection of macrophages, neurotoxic proteins, inflammatory factors produced by activated non-neuronal cells, in particularly microglia cells and astrocytes, in the brain of HIV-1-infected individuals are involved in neuronal apoptosis in HIV-1 infection. Therefore, prevention or treatment of neuronal injury in HIV-1 infection should include strategies to combat these factors. HAART has reduced the incidence of severe forms of HIV-1-associated neurologic disorders such as HAD, but with longer life span, the prevalence of milder forms of neurologic manifestations such as HAND appears to be increasing. One obstacle to control brain HIV-1 infection is the fact that antiretroviral drugs in the HAART regimens generally have a poor penetration into the CNS[112]. Thus, novel strategies to improve the delivery of these drugs to the CNS are urgently needed. Studies have shown that some host factors may affect the intracellular drug concentration leading to the inability of drug regimens to inhibit HIV-1 replication in cells[113]. For example, the ATP-binding cassette transporter proteins, such as P-glycoprotein and multidrug resistance-associated proteins (MRPs), affect efflux of nucleoside reverse transcriptase inhibitors (NRTIs) and protease inhibitors (PIs)[113]. Inhibition of P-glycoprotein or MRPs increases uptake of saquinavir, a protease inhibitor, in the mouse brain[114]. Therefore, targeting these transporter proteins may improve delivery of HAART regimens to the CNS.

Neurotoxic factors released from HIV-1-infected and/or activated macrophages/microglia play a key role in the neuronal apoptosis in HIV-1 infection. Therapies targeting activation of macrophages/microglia can potentially affect neuronal injury. Minocycline, a lipid soluble tetracycline antibiotic that has putative effects on immune system cells, has been proposed as a potential conjunctive therapy for HIV-1 associated neurologic disorders[115]. Minocycline has been shown to effectively cross the BBB into the CNS parenchyma to inhibit activation, proliferation, and viral replication of macrophages/microglia and lymphocytes *in vitro*[116-120], resulting in reduction of the production of immune activators by these cells and neurons as well[116,121,122]. *In vivo* studies in SIV-infected pigtailed macaques showed that minocycline reduced plasma virus, the pro-inflammatory monocyte chemoattractant protein 1 (MCP-1)/CCL2, and viral DNA in the CNS[115,117]. The mechanisms for these effects include inhibition of immune cell activation, down-regulation of CD16 expression on the surface of monocytes, reduction of monocyte/macrophage and infected cell traffic, and suppression of inflammatory responses[115].

Lexipafant (an antagonist of platelet-activating factor or PAF), prinomastat (an inhibitor of matrix metalloprotease or MMP), inhibitors of TNF-α, and antioxidants have also been proposed as therapeutic drugs for HIV-1 associated neurologic disorders[25]. Some of these inhibitors of neurotoxic factors have been used to prevent or treat the pathogenesis of HAD and HIV-1-related neurodegeneration. One compound that has yielded interesting results is CPI-1189 in the treatment of HAD[123]. CPI-1189 ameliorates TNF-α toxicity by increasing activation of ERK (extracellular signal-regulated kinase)-MAP kinase[124]. CPI-1189 also attenuates toxicity of macrophage culture obtained from HAD patients in the presence of quinolinic acid and gp120[124]. However, a great deal of additional work is still necessary to determine the true effectiveness of any of these therapeutic inhibitors.

Neuronal apoptosis in HIV-1 infection involves activation of NMDA receptors, alterations in Ca^{2+} homeostasis and increase of oxidative stress. MK801 (a NMDA receptor antagonist) or 7-nitroindazole (a NOS specific inhibitor) reduces gp120-induced neuronal apoptosis in the neocortex of rat[125]. Nimodipine, a voltage-dependent Ca^{2+} channel antagonist, significantly decreased the rise in intracellular Ca^{2+} in neurons, but not in astrocytes[123]. A phase I/II trial of nimodipine for HIV-1-related neurologic complications was conducted and the results showed that nimodipine was safe and tolerated by HIV-1-infected subjects with cognitive impairment[126]. Nimodipine has been tested as an adjuvant agent to HAART to improve neuropsychological performance of HIV-1-infected individuals[126]. In addition, a trend toward stabilization in peripheral neuropathy was observed in nimodipine-treated patients[126].

Inhibitors of proapoptotic factors and caspase-3, -8 and -9 have also been tested to treat HIV-1-associated neurologic disorders. Cultured rat cerebrocortical cells exposed to HIV-1 gp120 undergo activation of two upstream caspases including caspase-8 and caspase-9. Pretreatment of these neurons with pan-caspase inhibitor (zVAD-fmk), caspase-3 peptide inhibitor (DEVD-fmk), caspase-8 inhibitor (IETD-fmk), or caspase-9 inhibitor (LEHD-fmk) prevents gp120-induced neuronal apoptosis[127]. Specific inhibitors of both the Fas/TNF-α/death receptor pathway and the mitochondrial caspase pathway also prevent gp120-induced neuronal apoptosis. These data suggest that pharmacologic interventions aimed at the caspase enzyme pathways may be beneficial for the prevention or treatment of HAD or HIV-1-associated pain.

Our research group has recently reported that HIV-1 Tat protein directly causes hyperexcitability and apoptosis of DRG neurons, probably by inhibiting Cdk5 kinase activity and protein expression[62]. Tat-mediated hyperexcitability of DRG neurons may play a key role in the initiation of HIV-1-associated pain in patients at the early stages of the viral infection. HIV-1 Tat and other proteins induce apoptosis of DRG neurons, which may be involved in HIV-1-associated pain at all stages of viral infection. Our findings have potentially important implications for developing therapeutic strategies to prevent or treat HIV-1-assciated pain. For example, early initiation of HAART and/or neutralization of HIV-1 protein neurotoxicity may significantly prevent or delay HIV-1-associated pain. A recent study using a SIV-macaque model of HAART demonstrates that early initiation of HAART results in a dramatic reduction of viral RNA levels in plasma, CSF and brain[128], suggesting that early initiation of HAART can reduce or delay neurological complications including pain in HIV-1-infected patients. However, pain is often under-assessed and undertreated in people with HIV-1/AIDS illness, and the pain etiologies that are the key for pain management and improvement of the quality of life have been largely unexplored. In addition, like cancer pain, HIV-1/AIDS-associated pain tends to be of more than one type, to involve more than one location, and to increase in intensity as disease progression. Therefore, more comprehensive studies are urgently needed to investigate the current global burden of pain, pain types and origins, and the spectrum of neurological injuries in HIV-1/AIDS patients.

Author details

Mona Desai and Qigui Yu
Center for AIDS Research, Department of Microbiology and Immunology,
Indiana University School of Medicine, Indianapolis, Indiana, USA
Division of Infectious Diseases, Department of Medicine, Indiana University School of Medicine,
Indianapolis, Indiana, USA

Daniel Byrd
Center for AIDS Research, Department of Microbiology and Immunology,
Indiana University School of Medicine, Indianapolis, Indiana, USA

Ningjie Hu
Zhejiang Provincial Key Laboratory for Technology & Application of Model Organisms,
Wenzhou Medical College, University Park, Wenzhou, China

7. References

[1] Dube, B., Benton, T., Cruess, D. G. & Evans, D. L. Neuropsychiatric manifestations of HIV infection and AIDS. *Journal of psychiatry & neuroscience : JPN* 30, 237-246 (2005).

[2] Gray, G. & Berger, P. Pain in women with HIV/AIDS. *Pain* 132 Suppl 1, S13-21, doi:S0304-3959(07)00576-3 [pii]10.1016/j.pain.2007.10.009 (2007).

[3] Hughes, A. M. & Pasero, C. HIV-related pain. *Am J Nurs* 99, 20 (1999).

[4] Hirschfeld, S. Pain as a complication of HIV disease. *AIDS Patient Care STDS* 12, 91-108 (1998).

[5] Hirschfeld, S. & Morris, B. K. Review: pain associated with HIV infection. *Pediatr AIDS HIV Infect* 6, 63-74 (1995).

[6] Edmunds-Oguokiri, T. Pain management in HIV/AIDS: an update for clinicians. *HIV Clin* 19, 7-10 (2007).

[7] Verma, S., Estanislao, L. & Simpson, D. HIV-associated neuropathic pain: epidemiology, pathophysiology and management. *CNS Drugs* 19, 325-334, doi:1945 [pii] (2005).

[8] Del Borgo, C. *et al.* Multidimensional aspects of pain in HIV-infected individuals. *AIDS Patient Care STDS* 15, 95-102, doi:10.1089/108729101300003690 (2001).

[9] Dixon, P. & Higginson, I. AIDS and cancer pain treated with slow release morphine. *Postgrad Med J* 67 Suppl 2, S92-94 (1991).

[10] Lebovits, A. H. *et al.* The prevalence and management of pain in patients with AIDS: a review of 134 cases. *Clin J Pain* 5, 245-248 (1989).

[11] Singer, E. J. *et al.* Painful symptoms reported by ambulatory HIV-infected men in a longitudinal study. *Pain* 54, 15-19, doi:0304-3959(93)90094-6 [pii] (1993).

[12] Penfold, J. & Clark, A. J. Pain syndromes in HIV infection. *Can J Anaesth* 39, 724-730 (1992).

[13] O'Neill, W. M. & Sherrard, J. S. Pain in human immunodeficiency virus disease: a review. *Pain* 54, 3-14, doi:0304-3959(93)90093-5 [pii] (1993).

[14] Breitbart, W. & Dibiase, L. Current perspectives on pain in AIDS. *Oncology (Williston Park)* 16, 964-968, 972; discussion 972, 977, 980, 982 (2002).

[15] Hewitt, D. J. *et al.* Pain syndromes and etiologies in ambulatory AIDS patients. *Pain* 70, 117-123 (1997).

[16] Norval, D. A. Symptoms and sites of pain experienced by AIDS patients. *S Afr Med J* 94, 450-454 (2004).

[17] Kay, E., Dinn, J. J. & Farrell, M. A. Neuropathologic findings in AIDS and human immunodeficiency virus infection--report on 30 patients. *Irish journal of medical science* 160, 393-398 (1991).

[18] Trujillo, J. R. *et al.* International NeuroAIDS: prospects of HIV-1 associated neurological complications. *Cell research* 15, 962-969, doi:10.1038/sj.cr.7290374 (2005).

[19] Joska, J. A., Hoare, J., Stein, D. J. & Flisher, A. J. The neurobiology of HIV dementia: implications for practice in South Africa. *African journal of psychiatry* 14, 17-22 (2011).

[20] Woods, S. P., Moore, D. J., Weber, E. & Grant, I. Cognitive neuropsychology of HIV-associated neurocognitive disorders. *Neuropsychology review* 19, 152-168, doi:10.1007/s11065-009-9102-5 (2009).

[21] Mazus, A. I., Leven, II, Vinogradov, D. L., Chigrinets, O. V. & Dukhanina, I. V. [Neurologic manifestations of HIV infection]. *Klinicheskaia meditsina* 87, 59-60 (2009).

[22] Mouzat, L. [HIV and neurologic manifestations]. *Revue de l'infirmiere*, 17-18 (2006).

[23] Robertson, K. *et al.* A multinational study of neurological performance in antiretroviral therapy-naive HIV-1-infected persons in diverse resource-constrained settings. *Journal of neurovirology* 17, 438-447, doi:10.1007/s13365-011-0044-3 (2011).

[24] Price, R. W. *et al.* The brain in AIDS: central nervous system HIV-1 infection and AIDS dementia complex. *Science* 239, 586-592 (1988).

[25] Jones, G. & Power, C. Regulation of neural cell survival by HIV-1 infection. *Neurobiology of disease* 21, 1-17, doi:10.1016/j.nbd.2005.07.018 (2006).

[26] von Giesen, H. J., Koller, H., Hefter, H. & Arendt, G. Central and peripheral nervous system functions are independently disturbed in HIV-1 infected patients. *Journal of neurology* 249, 754-758, doi:10.1007/s00415-002-0707-3 (2002).

[27] Petito, C. K. & Roberts, B. Evidence of apoptotic cell death in HIV encephalitis. *The American journal of pathology* 146, 1121-1130 (1995).

[28] Gelbard, H. A. *et al.* Apoptotic neurons in brains from paediatric patients with HIV-1 encephalitis and progressive encephalopathy. *Neuropathology and applied neurobiology* 21, 208-217 (1995).

[29] Adle-Biassette, H. *et al.* Neuronal apoptosis in HIV infection in adults. *Neuropathology and applied neurobiology* 21, 218-227 (1995).

[30] Shi, B. *et al.* Apoptosis induced by HIV-1 infection of the central nervous system. *The Journal of clinical investigation* 98, 1979-1990, doi:10.1172/JCI119002 (1996).

[31] An, S. F., Gray, F. & Scaravilli, F. Programmed cell death in brains of HIV-1-positive pre-AIDS patients. *Lancet* 346, 911-912 (1995).

[32] An, S. F. *et al.* Programmed cell death in brains of HIV-1-positive AIDS and pre-AIDS patients. *Acta neuropathologica* 91, 169-173 (1996).

[33] Breitbart, W. & Dibiase, L. Current perspectives on pain in AIDS. *Oncology (Williston Park)* 16, 818-829, 834-815 (2002).

[34] Bagetta, G. *et al.* The HIV-1 gp120 causes ultrastructural changes typical of apoptosis in the rat cerebral cortex. *Neuroreport* 7, 1722-1724 (1996).

[35] Corasaniti, M. T., Nistico, R., Costa, A., Rotiroti, D. & Bagetta, G. The HIV-1 envelope protein, gp120, causes neuronal apoptosis in the neocortex of the adult rat: a useful experimental model to study neuroaids. *Funct Neurol* 16, 31-38 (2001).

[36] Wallace, V. C. *et al.* Characterization of rodent models of HIV-gp120 and anti-retroviral-associated neuropathic pain. *Brain* 130, 2688-2702, doi:awm195 [pii]10.1093/brain/awm195 (2007).

[37] Wallace, V. C. *et al.* Pharmacological, behavioural and mechanistic analysis of HIV-1 gp120 induced painful neuropathy. *Pain* 133, 47-63, doi:S0304-3959(07)00101-7 [pii]10.1016/j.pain.2007.02.015 (2007).

[38] Navia, B. A., Jordan, B. D. & Price, R. W. The AIDS dementia complex: I. Clinical features. *Ann Neurol* 19, 517-524, doi:10.1002/ana.410190602 (1986).

[39] Navia, B. A., Cho, E. S., Petito, C. K. & Price, R. W. The AIDS dementia complex: II. Neuropathology. *Ann Neurol* 19, 525-535, doi:10.1002/ana.410190603 (1986).

[40] Ranki, A. *et al.* Abundant expression of HIV Nef and Rev proteins in brain astrocytes in vivo is associated with dementia. *AIDS* 9, 1001-1008 (1995).

[41] de la Monte, S. M., Ho, D. D., Schooley, R. T., Hirsch, M. S. & Richardson, E. P., Jr. Subacute encephalomyelitis of AIDS and its relation to HTLV-III infection. *Neurology* 37, 562-569 (1987).

[42] Li, W., Galey, D., Mattson, M. P. & Nath, A. Molecular and cellular mechanisms of neuronal cell death in HIV dementia. *Neurotox Res* 8, 119-134 (2005).

[43] Mattson, M. P., Haughey, N. J. & Nath, A. Cell death in HIV dementia. *Cell Death Differ* 12 Suppl 1, 893-904, doi:4401577 [pii]10.1038/sj.cdd.4401577 (2005).

[44] Center, R. J., Earl, P. L., Lebowitz, J., Schuck, P. & Moss, B. The human immunodeficiency virus type 1 gp120 V2 domain mediates gp41-independent intersubunit contacts. *J Virol* 74, 4448-4455 (2000).

[45] Minami, T. *et al.* Functional evidence for interaction between prostaglandin EP3 and kappa-opioid receptor pathways in tactile pain induced by human immunodeficiency virus type-1 (HIV-1) glycoprotein gp120. *Neuropharmacology* 45, 96-105, doi:S0028390803001333 [pii] (2003).

[46] Robinson, B., Li, Z. & Nath, A. Nucleoside reverse transcriptase inhibitors and human immunodeficiency virus proteins cause axonal injury in human dorsal root ganglia cultures. *J Neurovirol* 13, 160-167, doi:778723504 [pii]10.1080/13550280701200102 (2007).

[47] Oh, S. B. *et al.* Chemokines and glycoprotein120 produce pain hypersensitivity by directly exciting primary nociceptive neurons. *The Journal of neuroscience : the official journal of the Society for Neuroscience* 21, 5027-5035 (2001).

[48] Keswani, S. C. *et al.* Schwann cell chemokine receptors mediate HIV-1 gp120 toxicity to sensory neurons. *Ann Neurol* 54, 287-296, doi:10.1002/ana.10645 (2003).

[49] Acharjee, S. *et al.* HIV-1 viral protein R causes peripheral nervous system injury associated with in vivo neuropathic pain. *FASEB J* 24, 4343-4353, doi:fj.10-162313 [pii]10.1096/fj.10-162313 (2010).

[50] Hill, J. M., Mervis, R. F., Avidor, R., Moody, T. W. & Brenneman, D. E. HIV envelope protein-induced neuronal damage and retardation of behavioral development in rat neonates. *Brain research* 603, 222-233 (1993).

[51] Toggas, S. M. *et al.* Central nervous system damage produced by expression of the HIV-1 coat protein gp120 in transgenic mice. *Nature* 367, 188-193, doi:10.1038/367188a0 (1994).

[52] Thomas, F. P., Chalk, C., Lalonde, R., Robitaille, Y. & Jolicoeur, P. Expression of human immunodeficiency virus type 1 in the nervous system of transgenic mice leads to neurological disease. *Journal of virology* 68, 7099-7107 (1994).

[53] Kim, B. O. *et al.* Neuropathologies in transgenic mice expressing human immunodeficiency virus type 1 Tat protein under the regulation of the astrocyte-specific glial fibrillary acidic protein promoter and doxycycline. *The American journal of pathology* 162, 1693-1707, doi:10.1016/S0002-9440(10)64304-0 (2003).

[54] Jones, G. J. *et al.* HIV-1 Vpr causes neuronal apoptosis and in vivo neurodegeneration. *The Journal of neuroscience : the official journal of the Society for Neuroscience* 27, 3703-3711, doi:10.1523/JNEUROSCI.5522-06.2007 (2007).

[55] Jolicoeur, P. The CD4C/HIV(Nef)transgenic model of AIDS. *Current HIV research* 9, 524-530 (2011).

[56] Rahim, M. M., Chrobak, P., Hu, C., Hanna, Z. & Jolicoeur, P. Adult AIDS-like disease in a novel inducible human immunodeficiency virus type 1 Nef transgenic mouse model: CD4+ T-cell activation is Nef dependent and can occur in the absence of lymphophenia. *Journal of virology* 83, 11830-11846, doi:10.1128/JVI.01466-09 (2009).

[57] Dickie, P. Nef modulation of HIV type 1 gene expression and cytopathicity in tissues of HIV transgenic mice. *AIDS research and human retroviruses* 16, 777-790, doi:10.1089/088922200308774 (2000).

[58] Brenneman, D. E. *et al.* Neuronal cell killing by the envelope protein of HIV and its prevention by vasoactive intestinal peptide. *Nature* 335, 639-642, doi:10.1038/335639a0 (1988).

[59] Lannuzel, A., Lledo, P. M., Lamghitnia, H. O., Vincent, J. D. & Tardieu, M. HIV-1 envelope proteins gp120 and gp160 potentiate NMDA-induced [Ca2+]i increase, alter [Ca2+]i homeostasis and induce neurotoxicity in human embryonic neurons. *The European journal of neuroscience* 7, 2285-2293 (1995).

[60] Adamson, D. C. *et al.* Immunologic NO synthase: elevation in severe AIDS dementia and induction by HIV-1 gp41. *Science* 274, 1917-1921 (1996).

[61] Kruman, II, Nath, A. & Mattson, M. P. HIV-1 protein Tat induces apoptosis of hippocampal neurons by a mechanism involving caspase activation, calcium overload, and oxidative stress. *Experimental neurology* 154, 276-288, doi:10.1006/exnr.1998.6958 (1998).

[62] Chi, X. *et al.* Direct effects of HIV-1 Tat on excitability and survival of primary dorsal root ganglion neurons: possible contribution to HIV-1-associated pain. *PloS one* 6, e24412, doi:10.1371/journal.pone.0024412 (2011).

[63] Richardson, J. D. & Vasko, M. R. Cellular mechanisms of neurogenic inflammation. *J Pharmacol Exp Ther* 302, 839-845, doi:10.1124/jpet.102.032797 (2002).

[64] Evans, A. R., Vasko, M. R. & Nicol, G. D. The cAMP transduction cascade mediates the PGE2-induced inhibition of potassium currents in rat sensory neurones. *J Physiol* 516 (Pt 1), 163-178 (1999).

[65] Lopshire, J. C. & Nicol, G. D. The cAMP transduction cascade mediates the prostaglandin E2 enhancement of the capsaicin-elicited current in rat sensory neurons: whole-cell and single-channel studies. *J Neurosci* 18, 6081-6092 (1998).

[66] Sinkovics, J. G. Programmed cell death (apoptosis): its virological and immunological connections (a review). *Acta microbiologica Hungarica* 38, 321-334 (1991).

[67] Elmore, S. Apoptosis: a review of programmed cell death. *Toxicologic pathology* 35, 495-516, doi:10.1080/01926230701320337 (2007).

[68] Wesselingh, S. L. *et al.* Intracerebral cytokine messenger RNA expression in acquired immunodeficiency syndrome dementia. *Annals of neurology* 33, 576-582, doi:10.1002/ana.410330604 (1993).

[69] Hsu, H., Xiong, J. & Goeddel, D. V. The TNF receptor 1-associated protein TRADD signals cell death and NF-kappa B activation. *Cell* 81, 495-504 (1995).

[70] Cheng, B., Christakos, S. & Mattson, M. P. Tumor necrosis factors protect neurons against metabolic-excitotoxic insults and promote maintenance of calcium homeostasis. *Neuron* 12, 139-153 (1994).

[71] Yoshida, H. *et al.* Synergistic stimulation, by tumor necrosis factor-alpha and interferon-gamma, of fractalkine expression in human astrocytes. *Neuroscience letters* 303, 132-136 (2001).

[72] Cotter, R. *et al.* Fractalkine (CX3CL1) and brain inflammation: Implications for HIV-1-associated dementia. *Journal of neurovirology* 8, 585-598, doi:10.1080/13550280290100950 (2002).

[73] Ryan, L. A. *et al.* TNF-related apoptosis-inducing ligand mediates human neuronal apoptosis: links to HIV-1-associated dementia. *Journal of neuroimmunology* 148, 127-139, doi:10.1016/j.jneuroim.2003.11.019 (2004).

[74] Hesselgesser, J. *et al.* CD4-independent association between HIV-1 gp120 and CXCR4: functional chemokine receptors are expressed in human neurons. *Current biology : CB* 7, 112-121 (1997).

[75] Hesselgesser, J. *et al.* Neuronal apoptosis induced by HIV-1 gp120 and the chemokine SDF-1 alpha is mediated by the chemokine receptor CXCR4. *Current biology : CB* 8, 595-598 (1998).

[76] Song, L., Nath, A., Geiger, J. D., Moore, A. & Hochman, S. Human immunodeficiency virus type 1 Tat protein directly activates neuronal N-methyl-D-aspartate receptors at an allosteric zinc-sensitive site. *Journal of neurovirology* 9, 399-403, doi:10.1080/13550280390201704 (2003).

[77] Li, W. *et al.* NMDA receptor activation by HIV-Tat protein is clade dependent. *The Journal of neuroscience : the official journal of the Society for Neuroscience* 28, 12190-12198, doi:10.1523/JNEUROSCI.3019-08.2008 (2008).

[78] Longordo, F. *et al.* The human immunodeficiency virus-1 protein transactivator of transcription up-regulates N-methyl-D-aspartate receptor function by acting at metabotropic glutamate receptor 1 receptors coexisting on human and rat brain noradrenergic neurones. *The Journal of pharmacology and experimental therapeutics* 317, 1097-1105, doi:10.1124/jpet.105.099630 (2006).

[79] Kim, H. J., Martemyanov, K. A. & Thayer, S. A. Human immunodeficiency virus protein Tat induces synapse loss via a reversible process that is distinct from cell death. *The Journal of neuroscience : the official journal of the Society for Neuroscience* 28, 12604-12613, doi:10.1523/JNEUROSCI.2958-08.2008 (2008).

[80] Eugenin, E. A. *et al.* HIV-tat induces formation of an LRP-PSD-95- NMDAR-nNOS complex that promotes apoptosis in neurons and astrocytes. *Proceedings of the National Academy of Sciences of the United States of America* 104, 3438-3443, doi:10.1073/pnas.0611699104 (2007).

[81] Nath, A. *et al.* Synergistic neurotoxicity by human immunodeficiency virus proteins Tat and gp120: protection by memantine. *Annals of neurology* 47, 186-194 (2000).

[82] Zuo, Y. *et al.* HIV-1 genes vpr and nef synergistically damage podocytes, leading to glomerulosclerosis. *J Am Soc Nephrol* 17, 2832-2843, doi:ASN.2005080878 [pii]10.1681/ASN.2005080878 (2006).

[83] Dhillon, N. *et al.* Molecular mechanism(s) involved in the synergistic induction of CXCL10 by human immunodeficiency virus type 1 Tat and interferon-gamma in macrophages. *J Neurovirol* 14, 196-204, doi:794315853 [pii]10.1080/13550280801993648 (2008).

[84] Theodore, S., Cass, W. A. & Maragos, W. F. Methamphetamine and human immunodeficiency virus protein Tat synergize to destroy dopaminergic terminals in the rat striatum. *Neuroscience* 137, 925-935, doi:S0306-4522(05)01159-0 [pii]10.1016/j.neuroscience.2005.10.056 (2006).

[85] Theodore, S., Stolberg, S., Cass, W. A. & Maragos, W. F. Human immunodeficiency virus-1 protein tat and methamphetamine interactions. *Ann N Y Acad Sci* 1074, 178-190, doi:1074/1/178 [pii]10.1196/annals.1369.018 (2006).

[86] Flora, G. *et al.* Proinflammatory synergism of ethanol and HIV-1 Tat protein in brain tissue. *Exp Neurol* 191, 2-12, doi:S0014488604002377 [pii]10.1016/j.expneurol.2004.06.007 (2005).

[87] Tsujimoto, Y. Role of Bcl-2 family proteins in apoptosis: apoptosomes or mitochondria? *Genes to cells : devoted to molecular & cellular mechanisms* 3, 697-707 (1998).

[88] Zou, H., Henzel, W. J., Liu, X., Lutschg, A. & Wang, X. Apaf-1, a human protein homologous to C. elegans CED-4, participates in cytochrome c-dependent activation of caspase-3. *Cell* 90, 405-413 (1997).

[89] Jacotot, E. *et al.* Control of mitochondrial membrane permeabilization by adenine nucleotide translocator interacting with HIV-1 viral protein rR and Bcl-2. *The Journal of experimental medicine* 193, 509-519 (2001).

[90] Jacotot, E. *et al.* The HIV-1 viral protein R induces apoptosis via a direct effect on the mitochondrial permeability transition pore. *The Journal of experimental medicine* 191, 33-46 (2000).

[91] Shostak, L. D. *et al.* Roles of p53 and caspases in the induction of cell cycle arrest and apoptosis by HIV-1 vpr. *Experimental cell research* 251, 156-165, doi:10.1006/excr.1999.4568 (1999).

[92] Lee, H. *et al.* A p53 Axis Regulates B Cell Receptor-Triggered, Innate Immune System-Driven B Cell Clonal Expansion. *J Immunol* 188, 6093-6108, doi:10.4049/jimmunol.1103037 (2012).

[93] Gougeon, M. L. Apoptosis as an HIV strategy to escape immune attack. *Nature reviews. Immunology* 3, 392-404, doi:10.1038/nri1087 (2003).

[94] Giaccia, A. J. & Kastan, M. B. The complexity of p53 modulation: emerging patterns from divergent signals. *Genes & development* 12, 2973-2983 (1998).

[95] Silva, C. *et al.* Growth hormone prevents human immunodeficiency virus-induced neuronal p53 expression. *Annals of neurology* 54, 605-614, doi:10.1002/ana.10729 (2003).

[96] Jordan-Sciutto, K. L., Wang, G., Murphy-Corb, M. & Wiley, C. A. Induction of cell-cycle regulators in simian immunodeficiency virus encephalitis. *The American journal of pathology* 157, 497-507, doi:10.1016/S0002-9440(10)64561-0 (2000).

[97] Garden, G. A. *et al.* HIV associated neurodegeneration requires p53 in neurons and microglia. *FASEB journal : official publication of the Federation of American Societies for Experimental Biology* 18, 1141-1143, doi:10.1096/fj.04-1676fje (2004).

[98] Korsmeyer, S. J. Regulators of cell death. *Trends in genetics : TIG* 11, 101-105, doi:10.1016/S0168-9525(00)89010-1 (1995).

[99] Kroemer, G. The proto-oncogene Bcl-2 and its role in regulating apoptosis. *Nature medicine* 3, 614-620 (1997).

[100] Krajewska, M. *et al.* Expression of Bcl-2 family member Bid in normal and malignant tissues. *Neoplasia* 4, 129-140, doi:10.1038/sj/neo/7900222 (2002).

[101] Ramirez, S. H. *et al.* Neurotrophins prevent HIV Tat-induced neuronal apoptosis via a nuclear factor-kappaB (NF-kappaB)-dependent mechanism. *Journal of neurochemistry* 78, 874-889 (2001).

[102] Haughey, N. J. & Mattson, M. P. Calcium dysregulation and neuronal apoptosis by the HIV-1 proteins Tat and gp120. *J Acquir Immune Defic Syndr* 31 Suppl 2, S55-61 (2002).

[103] Ushijima, H., Nishio, O., Klocking, R., Perovic, S. & Muller, W. E. Exposure to gp120 of HIV-1 induces an increased release of arachidonic acid in rat primary neuronal cell culture followed by NMDA receptor-mediated neurotoxicity. *The European journal of neuroscience* 7, 1353-1359 (1995).

[104] Kumar, A., Manna, S. K., Dhawan, S. & Aggarwal, B. B. HIV-Tat protein activates c-Jun N-terminal kinase and activator protein-1. *J Immunol* 161, 776-781 (1998).

[105] Milani, D. *et al.* Extracellular human immunodeficiency virus type-1 Tat protein activates phosphatidylinositol 3-kinase in PC12 neuronal cells. *The Journal of biological chemistry* 271, 22961-22964 (1996).

[106] Magnuson, D. S., Knudsen, B. E., Geiger, J. D., Brownstone, R. M. & Nath, A. Human immunodeficiency virus type 1 tat activates non-N-methyl-D-aspartate excitatory amino acid receptors and causes neurotoxicity. *Annals of neurology* 37, 373-380, doi:10.1002/ana.410370314 (1995).

[107] Haughey, N. J., Holden, C. P., Nath, A. & Geiger, J. D. Involvement of inositol 1,4,5-trisphosphate-regulated stores of intracellular calcium in calcium dysregulation and neuron cell death caused by HIV-1 protein tat. *Journal of neurochemistry* 73, 1363-1374 (1999).

[108] Mayne, M. *et al.* HIV-1 tat molecular diversity and induction of TNF-alpha: implications for HIV-induced neurological disease. *Neuroimmunomodulation* 5, 184-192 (1998).

[109] Nath, A., Conant, K., Chen, P., Scott, C. & Major, E. O. Transient exposure to HIV-1 Tat protein results in cytokine production in macrophages and astrocytes. A hit and run phenomenon. *The Journal of biological chemistry* 274, 17098-17102 (1999).

[110] Scala, G. *et al.* The expression of the interleukin 6 gene is induced by the human immunodeficiency virus 1 TAT protein. *The Journal of experimental medicine* 179, 961-971 (1994).

[111] Igney, F. H. & Krammer, P. H. Death and anti-death: tumour resistance to apoptosis. *Nature reviews. Cancer* 2, 277-288, doi:10.1038/nrc776 (2002).

[112] Enting, R. H. *et al.* Antiretroviral drugs and the central nervous system. *AIDS* 12, 1941-1955 (1998).

[113] Turriziani, O. & Antonelli, G. Host factors and efficacy of antiretroviral treatment. *The new microbiologica* 27, 63-69 (2004).

[114] Park, S. & Sinko, P. J. P-glycoprotein and mutlidrug resistance-associated proteins limit the brain uptake of saquinavir in mice. *The Journal of pharmacology and experimental therapeutics* 312, 1249-1256, doi:10.1124/jpet.104.076216 (2005).

[115] Campbell, J. H. *et al.* Minocycline inhibition of monocyte activation correlates with neuronal protection in SIV neuroAIDS. *PloS one* 6, e18688, doi:10.1371/journal.pone.0018688 (2011).

[116] Nikodemova, M., Watters, J. J., Jackson, S. J., Yang, S. K. & Duncan, I. D. Minocycline down-regulates MHC II expression in microglia and macrophages through inhibition of IRF-1 and protein kinase C (PKC)alpha/betaII. *The Journal of biological chemistry* 282, 15208-15216, doi:10.1074/jbc.M611907200 (2007).

[117] Zink, M. C. *et al.* Neuroprotective and anti-human immunodeficiency virus activity of minocycline. *JAMA: the journal of the American Medical Association* 293, 2003-2011, doi:10.1001/jama.293.16.2003 (2005).

[118] Tikka, T. M. & Koistinaho, J. E. Minocycline provides neuroprotection against N-methyl-D-aspartate neurotoxicity by inhibiting microglia. *J Immunol* 166, 7527-7533 (2001).

[119] Tikka, T., Fiebich, B. L., Goldsteins, G., Keinanen, R. & Koistinaho, J. Minocycline, a tetracycline derivative, is neuroprotective against excitotoxicity by inhibiting activation and proliferation of microglia. *The Journal of neuroscience : the official journal of the Society for Neuroscience* 21, 2580-2588 (2001).

[120] Si, Q. *et al.* A novel action of minocycline: inhibition of human immunodeficiency virus type 1 infection in microglia. *Journal of neurovirology* 10, 284-292, doi:10.1080/13550280490499533 (2004).

[121] Shan, S. *et al.* New evidences for fractalkine/CX3CL1 involved in substantia nigral microglial activation and behavioral changes in a rat model of Parkinson's disease. *Neurobiology of aging* 32, 443-458, doi:10.1016/j.neurobiolaging.2009.03.004 (2011).

[122] Zhao, C., Ling, Z., Newman, M. B., Bhatia, A. & Carvey, P. M. TNF-alpha knockout and minocycline treatment attenuates blood-brain barrier leakage in MPTP-treated mice. *Neurobiology of disease* 26, 36-46, doi:10.1016/j.nbd.2006.11.012 (2007).

[123] Wallace, D. R. HIV neurotoxicity: potential therapeutic interventions. *Journal of biomedicine & biotechnology* 2006, 65741, doi:10.1155/JBB/2006/65741 (2006).

[124] Pulliam, L. *et al.* CPI-1189 attenuates effects of suspected neurotoxins associated with AIDS dementia: a possible role for ERK activation. *Brain research* 893, 95-103 (2001).

[125] Bagetta, G. *et al.* Inducible nitric oxide synthase is involved in the mechanisms of cocaine enhanced neuronal apoptosis induced by HIV-1 gp120 in the neocortex of rat. *Neuroscience letters* 356, 183-186, doi:10.1016/j.neulet.2003.11.065 (2004).

[126] Navia, B. A. *et al.* A phase I/II trial of nimodipine for HIV-related neurologic complications. *Neurology* 51, 221-228 (1998).

[127] Garden, G. A. *et al.* Caspase cascades in human immunodeficiency virus-associated neurodegeneration. *The Journal of neuroscience : the official journal of the Society for Neuroscience* 22, 4015-4024, doi:20026351 (2002).

[128] Graham, D. R. *et al.* Initiation of HAART during acute simian immunodeficiency virus infection rapidly controls virus replication in the CNS by enhancing immune activity and preserving protective immune responses. *Journal of neurovirology* 17, 120-130, doi:10.1007/s13365-010-0005-2 (2011).

Chapter 4

Translational Control in Tumour Progression and Drug Resistance

Carmen Sanges, Nunzia Migliaccio, Paolo Arcari and Annalisa Lamberti

Additional information is available at the end of the chapter

1. Introduction

Protein biosynthesis is a multi-step process that starts with the transcription of nuclear DNA, depository of genetic information, into messenger RNA (mRNA) that is used as template for the following polypeptide chain synthesis, also known as translation. Each step of this essential process is highly controlled in order to modulate any specific protein requirement of the cell in response to different stimuli and cellular events. This regulatory process is called translational control. Deregulation of the core signalling network in translational control, the phosphatidyl inositol trisphosphate kinase (PI3K), Protein Kinase B (PKB or Akt), mammalian target of rapamycin (mTOR) and RAS mitogen-activated protein kinase (MAPK)/MAPK-interacting Kinases (MNK) pathways, frequently occurs in human cancers and leads to aberrant modulation of mRNA translation. However, investigations on the contribution of these two pathways to translational regulation led to the interesting finding that translation factors are also substrate of signalling molecules. Post-translational modifications, including cleavage and phosphorylation, usually affect translational factors activity in protein biosynthesis; on the other hand, direct interaction of translational components with signalling mediators can either activate the pathway in which the mediator is involved or redirect translation factors to other activities, such as cytoskeletal rearrangements. These findings shed light on new functions of translation factors, different from their canonical role in protein synthesis. Taken together, these new functions are an intriguing step forward to the discovery of molecular mechanisms at the base of cellular response during "special" conditions such as cancer and drug resistance.

2. Translational machinery

Protein biosynthesis is a process present in all organisms, eukaryotes and prokaryotes, sharing similar mechanisms. In particular, translation starts at the ribosome and involves four different stages: initiation, elongation, termination and recycling [1]. All of these stages are tightly controlled by specific translation factors. Many of these factors are GTPases that are activated upon binding to the ribosome on a site called the GTPase-activating centre (GAC) [2]. In eukaryotes, at the initiation point, the 40S ribosomal subunit, carrying the eukaryotic initiation factor 3 (eIF3), is bound by a ternary complex consisting of the eukaryotic initiation factor 2 (eIF2), GTP and methionyl initiator tRNA (Met-tRNAi), to form the 43S preinitation complex (Fig. 1). The recruitment of mRNA is due to the the eukaryotic initiation factor 4F (eIF4F) complex formed by the cap-binding protein eukaryotic initiation factor 4E (eIF4E), the scaffold protein eukaryotic initiation factor 4G (eIF4G) and the RNA helicase eukaryotic initiation factor 4A (eIF4A), stimulated by the accessory factor eukaryotic initiation factor 4B (eIF4B). 4E binding proteins (4E-BPs) can compete with eIF4G for binding eIF4E thus inhibiting the association with 5' mRNA cap structures. The eIF4F complex, together with the poly(A)-binding protein (PABP), is able to recognise the 5′-terminal cap or the 3′-terminal poly(A) tract of mRNA and to transfer it to the 43S complex, resulting now in the 48S complex. Once the first AUG has been recognized, the pre-initiation complex formed by the initiation factors enables the binding of the 60S ribosomal subunit to the 40S ribosomal subunit to form the 80S initiation complex, via GTP hydrolysis of eIF2-GTP mediated by the eukaryotic initiation factor 5A (eIF5A) [3]. Many virus infections and stresses can induce a switch from a cap-dependent to a cap-independent initiation of translation. In this case the eIF2 ternary complex binds to an internal ribosome entry site (IRES) present on the mRNA 5′ untranslated region (5′ UTR), driving the translation directly to the 60S association phase [4]. In both cases the complex is now ready to receive the first elongator tRNA and to start with the elongation stage of the biosynthesis.

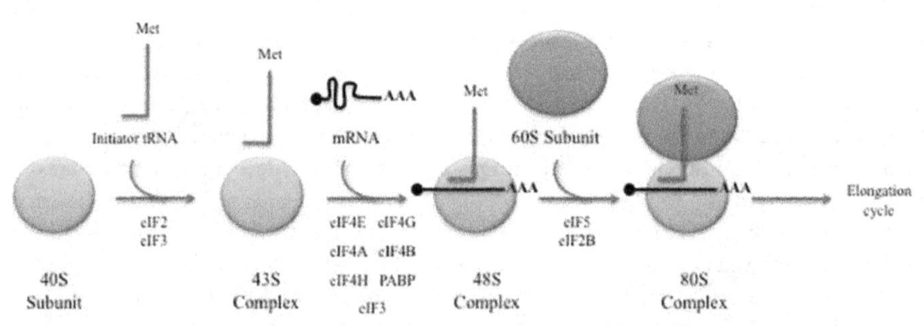

Figure 1. Eukaryotic translation initiation pathway. The initiation of translation in eukaryotes starts with the dissociation of the 40S ribosomal subunit from the 80S subunit probably promoted by eIF3 and eIF1A. Met-tRNA, eIF2 and GTP form a ternary complex that binds to the 40S ribosomal subunit to form the 43 S preinitiation complex. The eIF4 factors plus poly(A)-binding protein (PABP) recognize the mRNA [5′-terminal cap or 3′-terminal poly(A)] and transfer it to the 43 S initiation complex to form the 48 S initiation complex. After the recognition of the first initiation codon by eIF4A, in the presence of eIF1 and eIF1A, eIF5 stimulates GTP hydrolysis by eIF2 and the subsequent replacement of the initiation factors bound to 43S by the 60 S subunit to form the 80 S initiation complex. The released eIF2·GDP is recycled to eIF2·GTP by the GEF eIF2B. Adapted from Rhoads R.E. et al. [3].

The elongation phase of protein synthesis is a cyclic process consisting of basic steps repeated until the entire coding sequence of the mRNA is translated. In higher eukaryotes the elongation factors work always as a complex; for instance, the elongation factor 1 complex consists of eEF1A and the three subunit (β, γ, δ) of the elongation factor 1B (eEF1B) [5]. During elongation, a GTP-bound eEF1A transports the new aminoacyl-tRNA (aa-tRNA), as a ternary complex (Fig.2), to the empty A site of the 80S initiation complex [6]. In particular, eEF1A•GTP protects the aa-tRNA against hydrolysis and assists the ribosome in making a correct interaction between the current codon on the mRNA and the anticodon of the transported aa-tRNA [7]. Such a decoding event triggers the ribosome to induce GTP hydrolysis on eEF1A [8] and leads to its major conformational change that causes the release of aa-tRNA and the accommodation of the 3′ end in the peptidyl transferase (PT) centre on the 60S subunit, followed by peptide-bond formation [9]. In parallel, the inactive GDP-bound eEF1A (eEF1A•GDP) is released from the ribosome. This inactive form of eEF1A, unable to bind another aa-tRNA, is recycled to the active form (eEF1A•GTP) by exchange factor eEF1B. eEF1B consists of three subunits and works as a guanine nucleotide-exchange factor (GEF) for eEF1A [10]. In the following step of elongation, elongation factor 2 (eEF2) catalyzes the translocation of A and P site tRNAs to the P and E sites respectively, as well as movement of the mRNA by exactly one codon to allow a new round of elongation [11]. A-site-bound aa-tRNA reacts with P-site-bound pept-tRNA (peptidyl-tRNA) to form a peptide bond, resulting in deacylated tRNA in the P site and pept-tRNA prolonged by one amino acid in the A site.

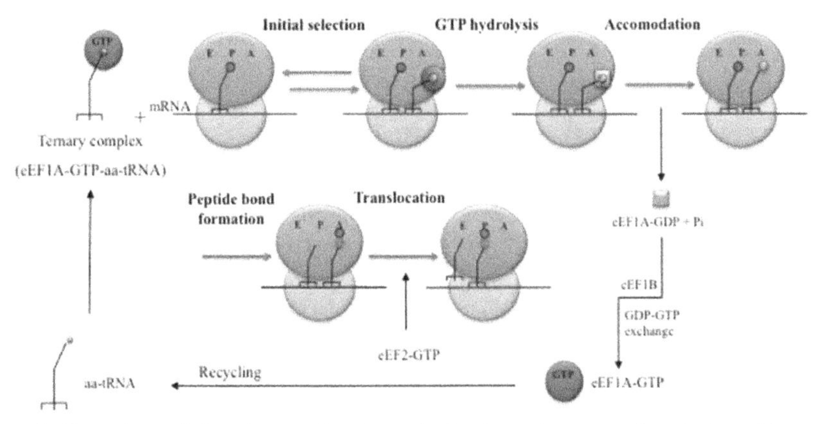

Figure 2. Elongation cycle in eukaryots. Crucial step in the elongation stage is the selection of the correct aa-tRNA, based on codon-anticodon interaction. The ribosome supervises this point during an initial selection of the ternary complex and during proofreading of the aa-tRNA. The initial selection utilizes the ability of cognate tRNA to stimulate the GTPase activity of eEF1A much faster than non-cognate and near-cognate tRNA.

When the ribosome come across one of the stop codons, UAA, UAG or UGA, eukaryoyic release factor 1 (eRF1) is recruited to the ribosome to promote the release of the newly synthesized polypeptide. After termination, the ribosome dissociates into its constituent subunits and the mRNA and deacylated tRNA is released thus allowing the ribosome to be recycled by eukaryoyic release factor 3 (eRF3) in cooperation with the eukaryotic elongation factor 2 (eEF2) [1].

3. Signalling transduction and translational control

Since regulatory mechanisms of protein biosynthesis are essential for maintaining a proper cellular metabolism, it is not surprising that the translation process is closely regulated by the coordinated activity of multiple intracellular signalling pathways acting at the centre of translational control. In eukaryotes, the most important pathways regulating the translation apparatus are the PI3K/Akt/mTOR and the Ras-MAPK signalling cascades [5] that can be stimulated by nutrient, insulin, growth factors and energy status (Fig. 3). Activation of these pathways mediates modifications such as changes in the phosphorylation states of translation factors and specific RNA-binding proteins, resulting in a translational machinery's activation or inhibition. Given the complexity of these two pathways, already well reviewed by several groups (we highly recommend Sonenberg N. et al. [5] and Proud C.G. [12]), in this paragraph we will just give a brief introduction (summarized in Table 1) of the main regulatory proteins participating in translational control and their influence on protein synthesis.

The lipid kinase PI3K is an important signalling mediator and its activation produces an increase of phosphatidyl inositol 3,4-biphosphate that activates downstream effectors such as the protein kinase B (PKB) also named Akt. The action of PI3K is antagonized by the lipid phosphatase and tensin homologue deleted on chromosome 10 (PTEN) [13]. Mutations in PTEN, present in many human tumours, lead to a constitutive activated Akt and mTOR signalling [14]. In translational control, PI3K plays its regulatory role by the activation of Akt and the consequent suppression of glycogen synthetase kinase 3 (GSK3) and its inhibitory activity on eIF2B [15]. Moreover PI3K can modulate mTOR signalling through Akt.

mTOR is a member of the phosphoinositide 3-kinase-related kinase (PIKK) family and exhibits protein kinase activity; some but not all mTOR functions are specifically repressed by rapamycin [16]. Rapamycin forms a complex with the immunophilin FK506 binding protein-12 (FKBP12) that binds to the FKBP12-rapamycin binding (FRB) domain of mTOR and inhibits its kinase activity. In mammalian cells, two functionally distinct mTOR complexes exist: mTOR complex 1 (mTORC1), containing mTOR, Raptor, and LST8; and mTOR complex 2 (mTORC2), containing Rictor, LST8, and Sin1. mTORC2 regulation and function remain largely unknown, although this complex has been linked to cytoskeletal rearrangements and cell survival through Akt [17]. mTORC1, among its many functions, promotes protein translation through activation of the S6 kinases (S6Ks) and inhibition of the eukaryotic initiation factor 4E binding protein 1 (4E-BP1) [18]. mTORC1 signalling can be modulated by PI3K through Akt. In particular Akt negatively regulates the tuberous sclerosis complex 2 (TSC2) GAP activity on the mTORC1 complex. TSC2 is a GTPase-activating protein (GAP) for the small G-protein Rheb [19] and forms a dimeric complex with TSC1. Rheb is a G-protein that stimulates mTOR activity [20]. Akt-mediated phosphorylation of TSC2 leads to the inhibition of its GAP activity towards Rheb, allowing Rheb to accumulate in its GTP-bound state and leading to the activation of mTORC1. mTORC1 signalling, through activation of Akt, is also stimulated in case of loss of PTEN function. Interestingly, stimulation of mTORC1 resulting from constitutive Akt activation leads to a transformed phenotype. In fact, the proliferation of some tumour-derived cell lines (e.g. those lacking the tumour suppressor PTEN) can be inhibited by rapamycin, confirming a key role for signalling through mTORC1 [21]. mTORC1 is also linked to a range of other oncogenes or proto-oncogenes, including Ras, nuclear factor (NF) and the liver kinase B1 (LKB1) [22]. mTOR pathway can also positively modulates eEF2 activity by phosphorylating and suppressing the Ca2+/calmodulin-dependent kinase III (CaMKIII) ability to bind to calmodulin [23]. CaMKIII, also known as calcium/calmodulin-dependent eukaryotic elongation factor 2 kinase (eEF2K) is a specific calcium/calmodulin-dependent enzyme that regulates protein synthesis [24]. The only known substrate of this kinase is eEF2. eEF2K phosphorylates eEF2 on T56 and stops peptide elongation by decreasing the affinity of eEF2 for the ribosome. Thus, eEF2K-mediated phosphorylation acts as an internal negative regulator of eEF2 translational activity [25].

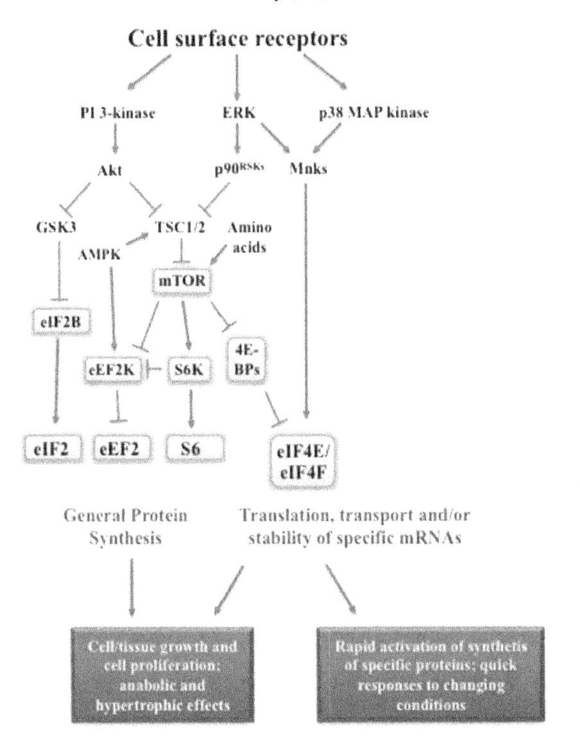

Figure 3. Translational control by signal transduction pathways. Activated signalling pathways modulate translation factors activity and mRNA-specific binding proteins regulating the rates of protein synthesis and translation and/or the stability of specific mRNAs. Adapted from Proud C.G. [12].

Signalling	Substrates	Effects
PI3K (via Akt and GSK3)	eIF2B (ε subunit)	Activity repression
	TSC2	mTORC1 upstream control
mTOR (via mTORC1)	4E-BP1	eIF4F complex positive control
	S6Ks	Activation and translation promotion
mTOR(unclear mediators)	eEF2 kinase	eEF2 activation
ERK (via p90rsk)	TSC2	mTORC1 upstream control
	eEF2 kinase	eEF2 activation
	eIF4B	Association with eIF3
p38 MAPK a/b (via MK2)	TSC2	mTORC1 upstream control
MAPK (via Mnks)	ARE-BPs(?)	mRNA stability
	eIF4G	Unclear
	eIF4E	Decrease of 5'-cap structure binding
	ARE-BPs(?)	mRNA stability
S6Ks	S6	Enanched binding to eIF3
	eIF4B	Unclear
	eEF2 kinase	Inhibition
	eEF1A	Stimulation
	eEF1B	Stimulation
AMPK	TSC2	mTORC1 inhibition
	eEF2 kinase	eEF2 inhibition
PKC	eEF1A	Increased elongation activity
	eEF1B	Increased aminoacylation activity
CK2	eEF1B	Inhibited interaction with eEF1A

Table 1. Signalling transduction in protein biosynthesis. Overview of the kinases involved in post-translational modifications on translation factors and related proteins and their effects on translational control. For more details and references see the text.

Classical MAPKs extracellular signal regulated kinase 1 and 2 (ERK1/2) and p38 MAPK a/b and c-Jun N-terminal kinase (JNK) pathways are the best understood MAPK signalling cascades in mammalian cells. The regulation of translational machinery through the MAPK signalling cascades is clearly connected with extraordinary events such as cancer and transformation. Each cascade involves downstream kinases that phosphorylate components of the translational machinery. In particular, ERK1/2 signalling activates p90 ribosomal s6 kinases ($p90^{rsk}$) that phosphorylate several translation factors or their regulators, including TSC2 [26], creating a link between ERK1/2 cascade and mTOR signalling [27]. Moreover p90^{rsk} phosphorylates eEF2 kinase, inhibiting eEF2 kinase activity [28], and eIF4B, promoting its association with eIF3 [29]. p38 MAPK a/b regulates the activation of the MAPK-activated protein kinase 2 (MK2) that mediates phosphorylation of TSC2 too [30]. Interestingly, MK2 is also able to control both the stability and the translation of some mRNAs, such as the tumour necrosis factor α (TNF-α) mRNA, containing at their 3'UTR a particular sequence called adenine/uridine-rich element (ARE), important for modulating the expression of specific proteins. It is likely that this regulating mechanism involves MK2-mediated modification of ARE binding proteins (ARE-BPs) [31]. MAPK signal-interacting kinases 1 and 2 (Mnk1 and 2) are both substrates for either ERK1/2 or p38 MAPKs a/b. During translational control, their activation leads to the phosphorylation of eIF4E [32] and eIF4G [5]. Phosphorylation of the former decreases eIF4E ability to bind 5'-cap structure while phosphorylation of the latter has unknown consequences.

Ribosomal S6 kinase (RSK) is involved at different levels in signal transduction. There are two subfamilies of RSK: p90^{rsk}, also known as MAPK-activated protein kinase-1 (MAPKAP-K1), and p70^{rsk}, also known as S6 Kinases (S6Ks). There are four variants of p90^{rsk} in humans (RSK 1-4) and two known mammalian homologues of S6Ks: S6K1 and S6K2. S6Ks are implicated in the positive regulation of cell growth and proliferation [33]. Once activated, S6Ks phosphorylate the S6 protein of the 40S ribosomal subunit and the translation initiation factor eIF4B to promote translation. S6Ks regulate mTOR through a negative feedback signalling pathway that affects insulin receptor substrate-1 (IRS-1). S6Ks was shown to directly phosphorylate IRS-1 to inhibit PI3K and Akt activation [34]. S6Ks activation decreases IRS-1 expression while rapamycin treatment restores IRS-1 expression [35]. Due to their major function in regulating translation, S6Ks are required also for cell growth and G1 cell cycle progression. Interestingly, S6K1 activation correlates with enhanced translation of a subset of mRNAs that contain a 5′-tract of oligopyrimidine (TOP mRNAs). These mRNAs encode for ribosomal proteins, elongation factors, the poly-A binding protein and other components of the translational machinery that become selectively translated by their TOP sequences in response to growth factors. However, S6Ks are not essential for the regulation of TOP mRNA translation [36].

AMP activated protein kinase (AMPK) is sensitive to the reduction of cellular content of AMP directly connected to that of ATP. In fact, in case of ATP decrease, cells need to slowdown protein synthesis in order to save energy; AMPK quickly reacts to this impair and negatively regulates mTORC1 via TSC2 phosphorylation [37]. Moreover AMPK can also modulate eEF2 activity by phosphorylating and activating eEF2K. The eEF2 phosphorylation mediated by eEF2K down-regulates eEF2 translational activity [38].

As above described, the best known mechanisms of translational control involve the initiation factors (2, 2B, 4B, 4E e 4G), the elongation factor 2 and the S6Ks. Less is known about the regulation of eEF1A and eEF1B even though several groups have studied the phosphorylation and regulation of these two factors [39]. Hereafter are summarized the current knowledge. The casein kinase 2 (CK2) phosphorylates the β subunit of eEF1B *in vitro* [40] leading to a decreased affinity of eEF1B toward eEF1A. Insulin or phorbol esters are able to enhance the phosphorylation of eEF1A and eEF1B *in vivo* [41]. Experimental results, including phosphopeptide-mapping, showed that insulin-stimulated multipotential S6 kinase was able to highly phosphorylate eEF1A and two subunits of the eEF1B complex (EF-1β and EF-1δ) from rabbit reticulocytes. However, phosphorylation of these proteins by S6 kinase *in vitro* resulted in a modest stimulation of their activity [42]. eEF1A and eEF1B are also substrates of protein kinase C (PKC) *in vitro* and this may explain the ability of phorbol esters (which activate several PKC isoforms) to increase the phosphorylation of these proteins *in vivo*. More precisely, PKCδ phosphorylates eEF1A at Threonine 431 (based on murine sequence) [43]. Phorbol esters also increase the phosphorylation of the valyl-tRNA synthetase that associates with eEF1A/B. The available evidence suggests that phosphorylation of eEF1A/B and of valyl-tRNA synthetase by PKC increases their activities in translation elongation and aminoacylation, respectively. The increased activity of eEF1A/B appears to result from enhanced GEF activity [44].

3.1. Translational control in apoptosis and tumour therapy

Generally, decrease in protein synthesis is an important adaptive mechanism that allows the cell to conserve or direct energy to other cellular functions. For example, upon induction of apoptosis a drastic reduction of protein biosynthesis occurs that precedes the loss of cell viability and the irreversible commitment to cell death [45, 46]. In fact, prior to and during the pro-apoptotic signal several factors with translational activity are modified. These modifications mainly include a specific caspase activity, prevented by the cell-permeable caspase-inhibitor z.VAD.FMK, and changes in the translation factors phosphorylation rates. In particular, eIF4B as well as eIF4GI, eIF4GII and eIF3j subunit are substrates of caspase-3 that mediates their cleavage and degradation following several pro-apoptotic stimuli [47]. In addition, eIF2, eIF4E and small 4E-BPs are highly phosphorylated during apoptosis with an inhibitory effect on protein synthesis [48]. The strong repression of translation initiation factors should lead to a complete inhibition of protein biosynthesis; however, a certain quote of cellular mRNA contain in their 5' UTR an IRES region that allows a cap-independent translation. IRESs directed translation is relatively inefficient under physiological conditions which favour cap-dependent translation, whereas it functions when cap-dependent translation is compromised [49]. Thus it is not surprising that cellular genes containing IRESs in their mRNAs, usually code for proteins that are involved in different cellular processes, including apoptosis. For instance, cleaved fragments of eIF4GI are able during apoptosis to enhance IRES-translation of the apoptotic protease-activating factor1 (Apaf-1) [50]. Another interesting example of translational control during apoptosis is the eukaryotic initiation factor 5A (eIF5A) and its peculiar post-translational modifications. eIF5A activity is modulated by a series of modifications that trigger the formation of the unusual amino acid hypusine [N-(4-amino-2-hydroxybutyl) lysine]. Hypusine plays a key role in the regulation of eIF5A function, as only the hypusine-containing eIF5A form is active. Intracellular hypusine content measures also the activity of eIF5A, as hypusine is contained only in this factor. Reduction of eIF5A1 expression or inhibition of hypusine modification may cause induction or suppression of apoptosis, depending on the biological system [51].

Besides eIFs, also other translation factors are involved in pro-apoptotic signalling. For example, upon treatment with antitumoural agents, eEF2 was shown to be involved in the therapeutic mechanism of doxorubicin. Treatment of prostatic cancer cells (PC3) with doxorubicin suppresses protein synthesis by inhibition of the elongation phase and not the initiation phase. This effect is probably mediated by a kinase independent phosphorylation of eEF2. Furthermore, inhibition of elongation activity correlates with decreased expression of the anti-apoptotic cellular FLICE-like inhibitor protein (cFLIPS), XIAP and survivin, all characterized by a short half-life and anti-apoptotic activity. These events result in a sensitization of the cells to the tumour-necrosis-factor-related apoptosis-inducing ligand (TRAIL) promoting, therefore, the doxorubicin apoptotic phenotype. TRAIL is a member of the TNF family capable to induce apoptosis in a wide variety of cancer cells upon binding to pro-apoptotic receptors, whereas it has no effect on the majority of normal human cells tested. Translation was found significantly inhibited in NIH3T3 cells also during taxol-induced apoptosis mediated by calpain [52]. Taxol treatment strongly decreased eIF4G, eIF4E and 4E-BP1 expression levels. However, a specific inhibitor of calpain, MDL28170, prevented reduction of eIF4G, but not of eIF4E or 4E-BP1 levels and did not block taxol-induced translation inhibition. Conversely, taxol treatment increased eEF2 phosphorylation in a calpain-independent manner thus supporting a role for eEF2 in taxol-induced translation inhibition [53].

3.1.1. *Eukaryotic elongation factor 1A*

Mentioning the relationship between apoptosis and translation, one cannot fail to mention the peculiar and fascinating role of the eukaryotic elongation factor 1A (eEF1A). In fact, eEF1A is a GTP binding protein that plays a central role in the elongation cycle of protein biosynthesis however, several studies suggest that eEF1A displays additional roles in different cellular processes far from its canonical role [54, 55]. eEF1A forms complexes with other cellular components like tubulin and actin. It has been observed that eEF1A cross-links actin filaments and it is implicated in microtubule binding, bundling or severing. Indeed, eEF1A mutants alter actin cytoskeleton organization but not translation, indicating a direct role of eEF1A on cytoskeletal organization in vivo [56]. In addition, eEF1A is known to be involved in several cellular process, including embryogenesis, senescense, oncogenic transformation, cell proliferation and organization of cytoskeleton [57]. In higher vertebrates, eEF1A is present in two isoforms (eEF1A1 and eEF1A2) with a different expression patterns and encoded by distinct genes [58]. The near-ubiquitous form, eEF1A1, is expressed in all tissues throughout development but is absent in adult muscle and heart expressing eEF1A2 instead [58, 59]. eEF1A2 is also found in some other cell types including large motor neurons, islet cells in the pancreas and enteroendocrine cells in the gut [60, 61]. Despite sharing 92% sequence identity, paralogous human eEF1A1 and eEF1A2 have different functional profiles. They exhibit similar translation activities but have different relative affinities for GTP and GDP [62] and, surprisingly, eEF1A2 appears to show little or no affinity for the components of the guanine-nucleotide exchange factor (GEF) complex eEF1B in yeast-two-hybrid experiments [63]. Moreover, as recent structural studies suggest, the two eEF1A isoforms display different behavior as tyrosine phosphorylation substrate that could affect their interaction with different signalling molecules. In fact, while eEF1A1 is able to interact with adaptor proteins containing SH2 domains, eEF1A2 is instead able to bind both SH2 and SH3 protein containing domains [64] thus suggesting for eEF1A2 a greater involvement in phosphotyrosine-mediated signalling processes [65]. Interestingly, the two isoforms play opposite roles during apoptosis. eEF1A1 expression has a marked pro-apoptotic effect, whereas expression of eEF1A2 correlates with differentiation and works as an inhibitor of caspase-mediated apoptosis (see next paragraph). In particular, antisense eEF1A1 provides the cells with significant protection from cell death upon induction of apoptosis by serum deprivation, *vice versa* eEF1A1 over-expression leads to a faster rate of cell death [66]. Moreover, eEF1A1 protein levels undergo rapid increase upon treatment with lethal doses of H_2O_2; pre-treatment of rat heart myoblast cells H9c2(2-1) with transcriptional inhibitors fails to abolish the oxidant-induced increase in eEF1A1. Furthermore, eEF1A1 mRNA levels remain steady throughout H_2O_2 treatment, suggesting that the up-regulation of eEF1A1 is mediated post-transcriptionally. Transient depletion of eEF1A1 protects the cells against H_2O_2-mediated cytotoxicity in proportion to the degree of repression of eEF1A1 protein levels thus suggesting that up-regulation of eEF1A1 plays a role in expediting the execution of the apoptotic program in response to oxidative stress [67]. Interestingly, upon serum deprivation-induced apoptosis, eEF1A2 protein disappears and is replaced by eEF1A1 in dying myotubes. In addition, continuous expression of eEF1A2 protects differentiated myotubes from apoptosis by delaying their death thus suggesting a prosurvival function for eEF1A2 in skeletal muscle. In contrast, myotube death is accelerated by the introduction of the eEF1A1 homologues gene [68]. Investigations of eEF1A1 functional role related to apoptosis seems to be particulary promising as it may affect both protein synthesis and cytoskeletal organization, fundamental events during death signalling. A recent example of eEF1A1 involvement in apoptosis suggests that this protein mediates lipotoxic cell death through a mechanism independent from changes in the rates of protein synthesis. Since eEF1A1 plays an important role in remodelling microtubules and filamentous actin [56, 69, 70] and because the cytoskeleton undergoes dramatic changes during apoptosis and cell death, eEF1A1 may mediate cytoskeletal changes required to execute cell death programs in response to lipotoxic conditions [71].

3.2. Anti-apoptotic activity in cancer cells. Role of translation factors

Several factors of the translational machinery are involved at different levels in tumour progression with a strong support of the anti-apoptotic activity characteristic of cancer cells. In general, survival of most mammalian cells is dependent on extracellular signals that suppress programmed cell death. Recent studies have shown that survival factors prevent apoptosis trough the activation of PI3 kinase (PI3k) pathway [72] and its downstream effector, the protein-serine/threonine kinase Akt [73]. PI3k/Akt signalling acts upstream of mitochondria preventing the release of cytochrome c and subsequent activation of cytosolic caspases [74]. However, the targets of PI3k/Akt signalling that promote cell survival remain to be fully elucidated. Interestingly, expression of the non-phosphorylable eIF2B mutant prevents cytochrome *c* release upon inhibition of PI3k whereas, inhibition of translation with cycloheximide induced cytochrome *c* release. Regulation of translation resulting from phosphorylation of eIF2B thus appears to affect the apoptotic cascade upstream of mitochondria, most likely interfering with PI3k/Akt signalling [75]. The mTOR/eIF4F axis is also an important contributor to tumour maintenance and progression program in terms of anti-apoptotic activity. Suppression of mTOR activity and that of the downstream translation regulators, including eIF4E, delayes breast cancer progression, onset of associated pulmonary metastasis *in vivo* and breast cancer cell invasion and migration *in vitro*. eIF4E regulates the recruitment of mRNA to ribosomes, and thereby globally regulates cap-dependent protein synthesis. However, its over-expression contributes to malignancy by selectively enabling the translation of a limited pool of mRNAs that generally encode for proteins involved in cellular growth, angiogenesis, survival and malignancy. Translation of vascular endothelial growth factor (VEGF), matrix metallopeptidase 9 (MMP9) and cyclin D1 mRNAs, encoding for products associated with the metastatic phenotype, is indeed inhibited upon eIF4E suppression. Transgenic eIF4E-expressing mice show a marked increase in tumourigenesis by developing tumours of various histologies. Thus, eIF4E acts as an oncogene in vivo [76]. Moreover, over-expressed eIF4E prevents Myc-dependent apoptosis, at least in part, through a cyclin D1-dependent process [77] and in part by its ability to increase cellular levels of $BclX_L$, a key apoptotic antagonist [78].

Because high level of protein synthesis is one of the characteristics of cancer cells, the elongation cycle has recently gained much more attention in this field as it seems to be directly involved in cell survival. eEF2, a critical enzyme of the elongation cycle, has been investigated as a target for new therapies and as a potential contributor to the success of conventional therapies. Interestingly, eEF2 is highly expressed in lung adenocarcinoma (LADC), but not in the non-tumour lung tissue. High eEF2 expression correlates with a significantly higher incidence of early tumour recurrence, and a significantly bad prognosis. Silencing of eEF2 expression increases mitochondrial elongation, cellular autophagy and cisplatin sensitivity. Moreover, eEF2 was found sumoylated and this sumoylation correlates with drug resistance. In particular, sumoylation of eEF2 is essential for protein stability and cell survival against cisplatin in LADC cells. Taken together, these results suggest eEF2 as an anti-apoptotic marker in LADC [79]. eEF2 protein is also over-expressed in 92.9% of gastric and 91.7% of colorectal cancers with no mutations in any of the exons of the eEF2 gene. Over-expressed eEF2 significantly enhances the cell growth through promotion of G2/M progression in cell cycle, activating Akt and cdc2 (G2/M regulator), and inactivating eEF2 kinase (negative regulator of eEF2). Conversely, knockdown of eEF2 inhibits cancer cell growth and induces G2/M arrest. These results provide a novel linkage between translational elongation and cell cycle mechanisms [80]. The implication of eEF2K (CaMKIII) in cancer was suggested by the observation that this kinase is up-regulated in various types of tumours such as malignant glioma and breast cancer. Inhibition of eEF2K results in a decreased viability of tumour cells. eEF2K was previously demonstrated to phosphorylate and in turn down-regulate eEF2 activity. However, eEF2K is not only a negative regulator of protein synthesis but also a positive regulator of autophagy, under environmental or metabolic stresses. Similarly, aberrant activation of Akt promotes cell growth, survival and proliferation, and is associated with cancer development and progression. Akt represents an attractive target for therapeutic intervention against cancer. Akt inhibitors, such as the allosteric small molecule MK-2206, induce either apoptosis or autophagy. Interestingly, recent studies demonstrated that silencing of eEF2K, upon Akt inhibition, can blunt autophagy and augment apoptosis, thereby modulating the sensitivity of cancer cells to Akt inhibitors. Thus, targeting eEF2K reinforces the anti-tumour efficacy of Akt inhibitors, such as MK-2206, by promoting the switch from autophagy to apoptosis [81].

As widely demonstrated, eEF1A is also clearly connected with cancer progression and survival. In humans, eEF1A2 shows oncogenic properties when over-expressed; moreover, it is implicated in ovarian, breast, pancreatic, liver and lung cancer thus becoming one of the most intriguing putative oncogenes in the last decade [82]. Notably, eEF1A2 can either directly or indirectly activate the Akt signalling pathway. Previous studies assessed a direct interaction between eEF1A2 and phosphorylated Akt 1 and 2 (pAkt) in breast cancer. eEF1A2 regulates pAkt levels promoting cell survival, tumour progression and motility [83]. In mouse fibroblast cell line NIH3T3, eEF1A2 interacts with peroxiredoxin-I (Prdx-I), resulting in increased activation of Akt, reduced activation of caspases 3 and 8, and protection against apoptotic death [84]. Moreover, downregulation of eEF1A2 expression leads to decreased expression of pAkt1 and to less extent of pAkt2 and promotes apoptosis [85]. Thus, eEF1A2 interaction with pAkt represents an important mechanism for the regulation of Akt-dependent survival signalling pathways in cancer [85, 86].

4. Survival and chemotherapy failure

Many chemotherapeutic agents exert their cytotoxic effects through the induction of apoptosis however, despite the fact that many tumours initially respond to therapy, tumor cells can subsequently survive by gaining resistance to these treatments. Therefore, emergence of drug resistance during chemotherapy is a major cause of cancer relapse and consequent therapy failure. In the last years there is an increase evidence that also translation factors participate in the control of tumor chemoresistance with mechanisms not yet well understood. For example, the initiation factor 4E (eIF4E) that is overexpressed in many solid tumors, plays a role not only in cell growth and proliferation but also in the apoptotic response and in the acquisition of drug resistance [87]. In fact, eIF4E controls the translation on an increasing number of mRNAs encoding proteins with notable functions in all aspects of malignancy, including angiogenesis and invasiveness through the activation of the ras and phosphatidylinositol 3-kinase/AKT anti-apoptotic pathways [78, 88, 89]. These findings clearly suggest that eIF4E is a promising target for anticancer therapy.

Also translation elongation factors can be involved in the regulation of drug resistance. For instance, eEF2 is phosphorylated at Thr56 by eEF2K thus terminating peptide elongation by decreasing its affinity for the ribosome. eEF2K is up-regulated in several types of malignancies, including gliomas, and affects the sensitivity of cancer cells to treatment with the tumor necrosis factor-related apoptosis-inducing ligand (TRAIL). This ligand is considered a promising candidate as an anticancer agent based on its ability to trigger rapid apoptosis and its specific cytotoxicity in malignant cells by binding to the death receptors DR4 (TRAIL-RI) and DR5 (TRAIL-RII). Inhibition of eEF2K by RNA interference (RNAi) or by a pharmacological inhibitor (NH125) recovers sensitization of tumour cells to TRAIL-induced apoptosis through down-regulation of the anti-apoptotic protein, Bcl-XL [90]. These results indicate a possible therapeutic strategy for enhancing the efficacy of TRAIL against malignant cells by targeting eEF2 kinase.

Taken togheter, these studies suggest a role for translation factors and translational control signalling pathways in drug resistence and chemotherapy failure. The elucidation of basic molecular mechanisms leading to chemoresistance is an essential step in the development of new anti-neoplastic therapies.

4.1. Interferon alpha chemotherapy resistance. A new mechanism mediated by eEF1A

Since the advent of genetic engineering technology, recombinant IFNα has been largely employed in solid tumours treatments. To date, interferon therapy is used in combination with chemotherapy and radiation as a treatment for many cancers, and several clinical trials are currently ongoing. However, IFNα displays a limited activity, and several cancers are resistant to its anti-tumour function having developed mechanisms not completely elucidated yet. In human epidermoid cancer cells, IFNα induces growth inhibition and apoptosis most likely through the activation of caspase cascade mediated by JNK-1 and/or p38 MAPK activation and the mitochondrial pathway [91]. Furthermore, a concomitant reduction of the hypusinated eIF5A1 expression levels and eIF5A1 activity is also observed (Figure 4) [92]. These anti-proliferative and pro-apoptotic activities are all antagonized by the epidermal growth factor (EGF); notably, IFNα was found to increase the functional expression of the epidermal growth factor receptor (EGFR), participating itself in the EGF-mediated survival pathway. Moreover, the increase of EGFR leads to an hyperactivation of the Ras-dependent MAPK (Ras->Raf-1->Mek1->Erk-1/2) signalling further stimulated by the addition of EGF and with a prominent role in the antiapoptotic effects exerted by EGF. In particular, Raf-1 activity is increased by either EGF or IFNα and is potentiated after EGF addition. Raf-1, also known as C-Raf, is a member of the Raf kinase family of serine/threonine-specific protein kinases, composed by three members: Raf-A, B-Raf and C-Raf. It functions downstream of the Ras subfamily of membrane associated GTPases to which it binds directly. Once activated, C-Raf can phosphorylate and activate the protein kinases MEK1 and MEK2 that in turn phosphorylate to activate ERK1 and ERK2. Interestingly, C-Raf is known to exert both kinase-dependent and kinase-independent tumour-promoting functions in several cancers [93]. Our studies correlate its increased activity in human epidermoid cancer cells, during the survival response upon IFNα treatment, with the over-expression and post-translational modifications of eEF1A.

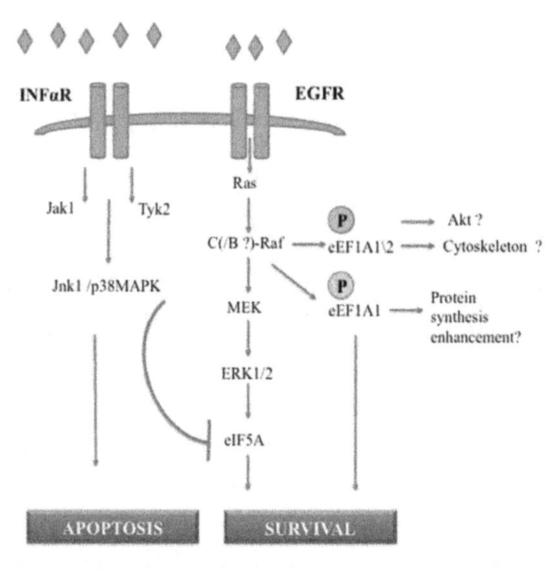

Figure 4. INFα and EGF signaling pathway. Schematic representation of the apoptotic and survival pathways mediated by INFαR and EGFR, respectively and their interaction with translation factors.

In particular, in human epidermoid lung cancer cells H1355, we found that upon treatment of the cells with INFα, both eEF1A1 and eEF1A2 protein levels increased but with a different degree since eEF-1A2 was the most up-regulated isoform. These data suggested that eEF1A2 increase was largely responsible for the upregulation of total eEF1A. To investigate the potential role of the increase in eEF1A protein levels, the apoptotic response to IFNα treatment was evaluated in H1355 cells in which eEF1A was down modulated by siRNA. In cells expressing low levels of eEF1A, the apoptotic cell death induced by IFNα was potentiated thus suggesting that eEF1A participate in these cells in the regulation of apoptosis. The increase of eEF1A levels mediated by INFα was also associated to phosphorylation of eEF1A on serine and threonine residues. These post-translation modifications have shown to be directly involved in the EGF-mediated survival response (see above) since the C-Raf inhibitor (BAY 43–9006) induces a decrease of eEF1A phopsphorylation. These data suggest the existence of an anti-apoptotic network between the translational factor 1A and the Ras-dependent signalling [94]. More specifically, we found that both eEF1A1 and eEF1A2 were singularly phosphorylated by B-Raf *in vitro*, whereas phosphorylation by C-Raf required the presence of both isoforms. Two new phosphorylation sites have been identified: T88 and S21. The former was specifically mediated by B-Raf on eEF1A1, whereas the latter was present on both eEF1A isoforms and mediated by both B- and C-Raf kinases. T88 phosphorylation was also identified on eEF1A1 expressed in proliferating COS 7 cells thus suggesting that this post-translational modification is isoform specific and probably due to structural differences between eEF1A1 and eEF1A2. PhosphoT88 might stabilize in vivo the elongation complex and improve then protein biosynthesis. In contrast, phosphorylation of S21, that belongs to the first GTP/GDP-binding consensus sequence (G14HVDSGKST in both eEF1A1 and eEF1A2), could potentially prevent the binding of eEF1A to guanine nucleotides thus switching eEF1A activity to different non-canonical functions. The finding that C-Raf required the presence of both eEF1A isoforms for its phosphorylation activity *in vivo* suggested that this switch might be regulated by the formation of a potential eEF1A heterodimer. Remarkably, eEF1A dimerization has been already described for *Tetrahymena* eEF1A. In this case, eEF1A bundles filamentous actin (F-actin) through dimer formation whereas eEF1A monomer do not [95]. A 3D model of the eEF1A1/eEF1A2 heterodimer was generated using as template the structure of yeast eEF1A (PDB ID: 1F60, chain A). As reported in Figure 5, the obtained docking model supports the possibility of a heterodimer between eEF1A1 and eEF1A2. In particular, this model shows that the M-domain of one isoform is in contact with the G-domain of the other and vice versa. The heterodimer formation somehow could induce a conformational change in one or in both eEF1A isoforms that allows the phosphorylation of S21. These speculations are confirmed by the finding that neither eEF1A1 nor eEF1A2, which were expressed in normal proliferating COS 7 cells, where the mitogenic cascade is particularly strong, showed any modifications of S21. However, phosphorylation on serine 21 might occur in tumour cells following the activation of a signal transduction pathway inducing tumourigenesis [96].

The discovered mechanism might gives a potential key resolution for the IFNα therapy resistance in lung cancer cells. Connection between translational control and mitogenic cascade open a new intriguing field of research in which studying the underlying mechanism of a potential Raf mediated regulation of eEF1A. The link between protein synthesis machinery and growth factor-elicited survival pathway represents an important molecular target to improve strategies based on apoptosis induction.

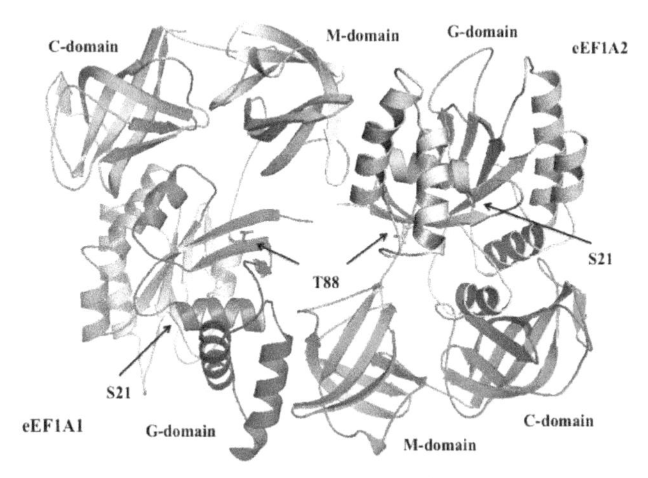

Figure 5. Imitation of a 3D model of an eEF1A1•eEF1A2 heterodimer. The heterodimer representation was obtained from the molecular docking pdb file (r-1.pdb) using PyMol software (DeLano Scientific LLC, San Carlos, CA, USA). In both eEF1A isoforms, the position of S21 and T88 are highlighted.

5. Conclusions

Events that cause alterations in protein synthesis and translational control have a particular role in the molecular mechanisms underlying cancer development and progression. Interestingly, several translation factors can be directly involved in signal transduction pathways, interact with oncogenes or probably act themselves as oncogenes. Alterations in translational control are also often associated with the molecular events participating in cell transformation, tumour development and progression, apoptosis induction or inhibition. All together, these evidences give translational control a central role in tumourigenesis and response to anti-neoplastic therapies. Thus, a deeper investigation of the specific changes in the translation apparatus for certain types of human cancers, in relation to their stage, grade, histopathology and exposure to standard anticancer therapies should be carried out. Understanding molecular alterations in translational control in each of these contexts can furnish possible indications to improve the use of therapeutic strategies in human cancer. In conclusion, we think that one possible way to improve tumour therapies is to better clarify specific cancer-associated changes in the translation machinery. This research could probably give the opportunity to develop selective anti-tumour translation inhibitors directed towards specific translational targets.

Author details

Carmen Sanges, Nunzia Migliaccio, Paolo Arcari and Annalisa Lamberti
Department of Biochemistry and Medical Biotechnologies, University of Naples Federico II, Naples, Italy
CEINGE, Advanced Biotechnologies scarl, Naples, Italy

Acknowledgement

This work was supported by funds from Programmi di Ricerca Scientifica di Rilevante Interesse Nazionale 2008 (2008BKRFBH_003) and PON Ricerca e Competitivita` 2007–2013 (PON01_02782). CS and PA were recipients of 'Deutsch-Italienisches Hochschulzentrum' (Progetto Vigoni 2008–2009). We are also grateful to Dr. Piero Ocone for valuable discussion and ideas.

6. References

[1] Noble CG, Song H. Structural studies of elongation and release factors. Cell. Mol. Life Science 2008;65:1335-1346.
[2] Ramakrishnan V. Ribosome structure and the mechanism of the translation. Cell 2002;108:557-572.
[3] Rhoads RE. Signal transduction Pathways that regulate eukaryotic protein synthesis. J. Biol. Chem. 1999;22:30337-30340.
[4] Komar AA, Hatzoglou M. Internal ribosome entry sites in cellular mRNAs: mystery of their existence. J. Biol. Chem. 2005;280:23425-23428.

[5] Sonenberg N, Hinnebusch AG. Regulation of translation initiation in eukaryotes: mechanisms and biological targets. Cell 2009;136:731–745.

[6] Stark H, Rodnina MV, Rinke-Appel J, Brimacombe R, Wintermeyer W, Heel M. Visualization of the elongation factor Tu on the E. Coli ribosome. Nature 1997;389:403-406.

[7] Pape T, Wintermeyer W, Rodnina M. Induced fit in initial selection and proofreading of aa-tRNA on the ribosome. EMBO 1999;18:3800-3807.

[8] Nilsson J, Nissen P. Elongation factor on the ribosome. Curr. Op. Struc. Biol. 2005;15:349-354.

[9] Nissen P, Hansen J. The structural basis of ribosome activity in peptide bond synthesis. Science 2000;289:920-930.

[10] Andersen GR, Valente L, Pedersen L, Kinzy TG, Nyborg J. Crystal structures of nucleotide exchange intermediates in the eEF1A–eEF1Ba complex. Nature Struc. Biol. 2001;8: 531-534.

[11] Rodnina MV, Savelsbergh A, Katunin VI, Wintermeyer W. Hydrolisi of GTP by elongation factor G drives tRNA movement on the ribosome. Nature 1997;385:37-41.

[12] Proud CG. Signalling to translation: how signal transduction pathways control the protein synthetic machinery. Biochem J. 2007;403:217-234.

[13] Maehama T, Dixon JE. The tumor suppressor, PTEN/MMAC1, dephosphorylates the lipid second messenger, phosphatidylinositol 3,4,5-trisphosphate. J Biol Chem. 1998;273: 13375–8.

[14] Zhang S, Yu D. PI(3)king apart PTEN's role in cancer. Clin Cancer Res. 2010;16:4325-30.

[15] Welsh GI, Miller CM, Loughlin, AJ, Price NT, Proud CG. Regulation of eukaryotic initiation factor eIF2B: glycogen synthase kinase-3 phosphorylates a conserved serine which undergoes dephosphorylation in response to insulin. FEBS Lett. 1998;421:125–130.

[16] Wullschleger S, Loewith R, Hall MN. TOR signalling in growth and metabolism. Cell 2006;124:471–484.

[17] Jacinto E, Loewith R, Schmidt A, Lin S, Rüegg MA, Hall A, Hall MN. Mammalian TOR complex 2 controls the actin cytoskeleton and is rapamycin insensitive. Nat. Cell Biol. 2004;6:1122-1128.

[18] Nojima H, Tokunaga C, Eguchi S, Oshiro N, Hidayat S, Yoshino K, Hara K, Tanaka J, Avruch J, Yonezawa K. The mTOR partner, raptor, binds the mTOR substrates, p70 S6 kinase and 4E-BP1, through their TOS (TOR signalling) motifs. J. Biol. Chem. 2003;278:15461–15464.

[19] Manning BD, Cantley LC. Rheb fills a GAP between TSC and TOR. Trends Biochem. Sci. 2003;28:573–576.

[20] Long X, Lin Y, Ortiz-Vega S, Yonezawa K, Avruch J. Rheb binds and regulates the mTOR kinase. Curr. Biol. 2005;15:702–713.

[21] Easton JB, Houghton PJ. mTOR and cancer therapy. Oncogene 2006;25:6436–6446.

[22] Sabatini DM. mTOR and cancer: insights into a complex relationship. Nat. Rev. Cancer 2006;6:729–734.

[23] Browne GJ, Proud CG. A novel mTOR-regulated phosphorylation site in elongation factor 2 kinase modulates the activity of the kinase and its binding to calmodulin. Mol Cell Biol. 2004;24:2986-2997.

[24] Ryazanov AG, Shestakova EA, Natapov PG. Phosphorylation of elongation factor 2 by EF-2 kinase affects rate of translation. Nature. 1988;334:170-173.
[25] Ryazanov AG. Ca2+/calmodulin-dependent phosphorylation of elongation factor 2. FEBS Lett. 1987;214:331-334.
[26] Rolfe M, McLeod LE, Pratt PF, and Proud CG. Activation of protein synthesis in cardiomyocytes by the hypertrophic agent phenylephrine requires the activation of ERK and involves phosphorylation of tuberous sclerosis complex 2 (TSC2). Biochem. J. 2005;388:973–984.
[27] Ma L, Chen Z, Erdjument-Bromage H, Tempst P, Pandolfi, PP. Phosphorylation and functional inactivation of TSC2 by Erk: implications for tuberous sclerosis and cancer pathogenesis. Cell 2005;121:179-193.
[28] Wang L, Proud CG. Regulation of the phosphorylation of elongation factor 2 by MEK-dependent signalling in adult rat cardiomyocytes. FEBS Lett. 2002;531:285–289.
[29] Shahbazian D, Roux PP, Mieulet V, Cohen MS, Raught B, Taunton J, Hershey JW, Blenis J, Pende M, Sonenberg N. The mTOR/PI3K and MAPK pathways converge on eIF4B to control its phosphorylation and activity. EMBO J. 2006;25:2781–2791.
[30] Li Y, Inoki K, Vacratsis P, Guan KL. The p38 and MK2 kinase cascade phosphorylates tuberin, the tuberous sclerosis 2 gene product, and enhances its interaction with 14-3-3. J. Biol. Chem. 2003;278:13663–13671.
[31] Hitti E, Iakovleva T, Brook M, Deppenmeier S, Gruber AD, Radzioch D, Clark AR, Blackshear PJ, Kotlyarov A, Gaestel M. Mitogen-activated protein kinase-activated protein kinase 2 regulates tumour necrosis factor mRNA stability and translation mainly by altering tristetraprolin expression, stability, and binding to adenine/uridine-rich element. Mol. Cell. Biol. 2006;26:2399–2407.
[32] Scheper GC, van Kollenburg B, Hu J, Luo Y, Goss DJ, Proud CG. Phosphorylation of eukaryotic initiation factor 4E markedly reduces its affinity for capped mRNA. J. Biol. Chem. 2002;277, 3303–3309.
[33] Thomas G. The S6 kinase signalling pathway in the control of development and growth. Biol Res. 2002;35:305-313.
[34] Harrington LS, Findlay GM, Gray A, Tolkacheva T, Wigfield S, Rebholz H, Barnett J, Leslie NR, Cheng S, Shepherd PR, Gout I, Downes CP, Lamb RF The TSC1-2 tumour suppressor controls insulin-PI3K signalling via regulation of IRS proteins. J. Cell Biol. 2004;166:213-223.
[35] Pende M, Um SH, Mieulet V, Sticker M, Goss VL, Mestan J, Mueller M, Fumagalli S, Kozma SC, Thomas G. S6K1(-/-)/S6K2(-/-) mice exhibit perinatal lethality and rapamycin-sensitive 5'-terminal oligopyrimidine mRNA translation and reveal a mitogen-activated protein kinase-dependent S6 kinase pathway. Mol. Cell Biol. 2004;24:3112-3124.
[36] Dufner A, Thomas G. Ribosomal S6 kinase signalling and the control of translation. Exp. Cell Res. 1999;253:100-109.
[37] Hardie DG, Scott JW, Pan DA, Hudson ER. Management of cellular energy by the AMP-activated protein kinase system. FEBS Lett. 2003;546:113–120.
[38] Browne GJ, Finn SG, Proud CG. Stimulation of the AMP-activated protein kinase leads to activation of eukaryotic elongation factor 2 kinase and to its phosphorylation at a novel site, serine 398. J. Biol. Chem. 2004;279:12220–12231.

[39] Browne GJ, Proud CG. Regulation of peptide-chain elongation in mammalian cells. Eur. J. Biochem. 2002;269:5360–5368.
[40] Sheu GT, Traugh JA. A structural model for elongation factor 1 (EF-1) and phosphorylation by protein kinase CKII. Mol. Cell. Biochem. 1999;191:181-186.
[41] Traugh JA. Insulin, phorbol ester and serum regulate the elongation phase of protein synthesis. Prog. Mol. Subcell. Biol. 2001;26:33-48.
[42] Chang YW, Traugh JA. Phosphorylation of elongation factor 1 and ribosomal protein S6 by multipotential S6 kinase and insulin stimulation of translational elongation. J. Biol Chem. 1997;272:28252-28257.
[43] Kielbassa K, Müller HJ, Meyer HE, Marks F, Gschwendt M. Protein kinase C delta-specific phosphorylation of the elongation factor eEF-alpha and an eEF-1 alpha peptide at threonine 431. J. Biol. Chem. 1995;270:6156-6162.
[44] Peters HI, Chang YW, Traugh JA. Phosphorylation of elongation factor 1 (EF-1) by protein kinase C stimulates GDP/GTP-exchange activity. Eur. J. Biochem. 1995;234:550-556.
[45] Zhou BB, Li H, Yuan J, Kirschner MW. Caspase-dependent activation of cyclin-dependent kinases during Fas-induced apoptosis in Jurkat cells. Proc. Natl. Acad. Sci. U S A. 1998;95:6785-6790.
[46] Scott CE, Adebodun F. 13C-NMR investigation of protein synthesis during apoptosis in human leukemic cell lines. J. Cell. Physiol. 1999;181:147-152.
[47] Morley SJ, Coldwell MJ, Clemens MJ. Initiation factor modifications in the preapoptotic phase. Cell Death and Differentiation 2005;12, 571–584.
[48] Holcik M, Sonenberg N. Translational control in stress and apoptosis. Nat. Rev. Mol. Cell Biol. 2005;6:318–327.
[49] Graber TE, Holcik M. Cap-independent regulation of gene expression in apoptosis. Mol Biosyst. 2007;3:825-834.
[50] Hanson PJ, Zhang HM, Hemida MG, Ye X, Qiu Y, Yang D. IRES-Dependent Translational Control during Virus-Induced Endoplasmic Reticulum Stress and Apoptosis. Front. Microbiol. Published online 2012 March 19. DOI: 10.3389/fmicb.2012.00092.
[51] Caraglia M, Park MH, Wolff EC, Marra M, Abbruzzese A. eIF5A isoforms and cancer: two brothers for two functions? Amino Acids. Published online 03 December 2011. DOI 10.1007/s00726-011-1182-x
[52] White SJ, Kasman LM, Kelly MM, Lu P, Spruill L, McDermott PJ, Voelkel-Johnson C. Doxorubicin generates a pro-apoptotic phenotype by phosphorylation of elongation factor 2. Free Radic. Biol. Med. 2007;43:1313-1321.
[53] Pineiro D, Gonzalez VM, Hernandez-Jimenez M, Salinas M, Martin ME. Translation regulation after taxol treatment in NIH3T3 cells involves the elongation factor (eEF)2. Exp. Cell. Res. 2007;313:3694–3706.
[54] Mateyak MK, Kinzy TG. eEF1A: thinking outside the ribosome. J. Biol. Chem. 2010;285:21209-21213.
[55] Ejiri S. Moonlighting functions of polypeptide elongation factor 1: from actin bundling to zinc finger protein R1-associated nuclear localization. Biosci. Biotechnol. Biochem. 2002;66:1-21.

[56] Gross SR, Kinzy TG. Translation elongation factor 1A is essential for regulation of the actin cytoskeleton and cell morphology Nature Structural & Molecular Biology 2005;12:772-778.
[57] Lamberti A, Caraglia M, Longo O, Marra M, Abbruzzese A, Arcari P. The translation elongation factor 1A in tumorigenesis, signal transduction and apoptosis: Review article. Amino Acids 2004;26:443–448.
[58] Lund A, Knudsen SM, Vissing H, Clark B, Tommerup N. Assignment of human elongation factor 1alpha genes: EEF1A maps to chromosome 6q14 and EEF1A2 to 20q13.3. Genomics 1996;36:359-361.
[59] Lee S, Francoeur AM, Liu S, Wang E. Tissue-specific expression in mammalian brain, heart, and muscle of S1, a member of the elongation factor-1 alpha gene family. J. Biol. Chem. 1992;267:24064-24068.
[60] Chambers DM, Peters J, Abbott CM. The lethal mutation of the mouse wasted (wst) is a deletion that abolishes expression of a tissue-specific isoform of translation elongation factor 1α, encoded by the Eef1a2 gene. Proc. Natl. Acad. Sci. USA. 1998;95:4463-4468.
[61] Newbery HJ, Loh DH, O'Donoghue JE, Tomlinson VA, Chau YY, Boyd JA, Bergmann JH, Brownstein D, Abbott CM. Translation elongation factor eEF1A2 is essential for post-weaning survival in mice. J Biol Chem. 2007;282:28951-28959.
[62] Kahns S, Lund A, Kristensen P, Knudsen CR, Clark BF, Cavallius J. The elongation factor 1 A-2 isoform from rabbit: Cloning of the cDNA and characterization of the protein. Nucleic Acids Res. 1998;26:1884-1890.
[63] Mansilla F, Friis I, Jadidi M, Nielsen KM, Clark BFC, Knudsen CR. Mapping the human translation elongation factor eEF1H complex using the yeast two-hybrid system. Biochem. J. 2002;365, 669-676.
[64] Schlessinger J. SH2/SH3 signaling proteins. Curr. Opin. Genet. Dev. 1994;4:25-30.
[65] Panasyuk G, Nemazanyy I, Filonenko V, Negrutskii B, El'skaya AV. A2 isoform of mammalian translation factor eEF1A displays increased tyrosine phosphorylation and ability to interact with different signalling molecules. Int. J. Biochem. Cell. Biol. 2008;40:63-71.
[66] Duttaroy A, Bourbeau D, Wang XL, Wang E. Apoptosis rate can be accelerated or decelerated by over-expression or reduction of the level of elongation factor-1 alpha. Exp. Cell. Res. 1998;238:168-176.
[67] Chen E, Proestou G, Bourbeau D, Wang E. Rapid up-regulation of peptide elongation factor EF-1alpha protein levels is an immediate early event during oxidative stress-induced apoptosis. Exp. Cell. Res. 2000;259:140-148.
[68] Ruest LB, Marcotte R, Wang E. Peptide elongation factor eEF1A-2/S1 expression in cultured differentiated myotubes and its protective effect against caspase-3-mediated apoptosis. J. Biol. Chem. 2002;277:5418-5425.
[69] Lamberti A, Sanges C, Longo O, Chambery A, Di Maro A, Parente A, Masullo M, Arcari P. Analysis of nickel-binding peptides in a human hepidermoid cancer cell line by Ni-NTA affinity chromatography and mass spectrometry. Protein Pept Lett. 2008;15:1126-1131.
[70] Pittman YR, Kandl K, Lewis M, Valente L, Kinzy TG. Coordination of eukaryotic translation elongation factor 1A (eEF1A) function in actin organization and translation

elongation by the guanine nucleotide exchange factor eEF1Balpha. J. Biol. Chem. 2009;284:4739-4747.

[71] Borradaile NM, Buhman KK, Listenberger LL, Magee CJ, Morimoto ET, Ory DS, Schaffer JE. A critical role for eukaryotic elongation factor 1A-1 in lipotoxic cell death. Mol. Biol. Cell. 2006;17:770-778.

[72] Yao R, Cooper GM. Requirement for phosphatidylinositol-3 kinase in the prevention of apoptosis by nerve growth factor. Science 1995;267:2003-2006.

[73] Datta, SR, Brunet A, Greenberg ME. Cellular survival: a play in three. Akts. Genes Dev. 1999;13:2905–2927.

[74] Kennedy SG, Kandel ES, Cross TK, Hay N. Akt/protein kinase B inhibits cell death by preventing the release of cytochrome c from mitochondria. Mol. Cell. Biol. 1999; 19:5800-5810.

[75] Pap M, Cooper GM. Role of Translation Initiation Factor 2B in Control of CellSurvival by the Phosphatidylinositol 3-Kinase/Akt/GlycogenSynthase Kinase 3β Signaling Pathway. Mol. Cell. Biol. 2002;22:578-586.

[76] Nasr Z, Robert F, Porco JA Jr, Muller WJ, Pelletier J. eIF4F suppression in breast cancer affects maintenance and progression. Oncogene. Published online 2012 Apr 9. DOI: 10.1038/onc.2012.105.

[77] Tan A, Bitterman P, Sonenberg N, Peterson M, Polunovsky V. Inhibition of Myc-dependent apoptosis by eukaryotic translation initiation factor 4E requires cyclin D1. Oncogene 2000;19:1437-1447.

[78] Wendel HG, De Stanchina E, Fridman JS, Malina A, Ray S, Kogan S, Cordon-Cardo C, Pelletier J, Lowe SW. Survival signalling by Akt and eIF4E in oncogenesis and cancer therapy. Nature 2004;428:332-337.

[79] Chen CY, Fang HY, Chiou SH, Yi SE, Huang CY, Chiang SF, Chang HW, Lin TY, Chiang IP, Chow KC. Sumoylation of eukaryotic elongation factor 2 is vital for protein stability and anti-apoptotic activity in lung adenocarcinoma cells. Cancer Sci. 2011;102:1582-1589.

[80] Nakamura J, Aoyagi S, Nanchi I, Nakatsuka S, Hirata E, Shibata S, Fukuda M, Yamamoto Y, Fukuda I, Tatsumi N, Ueda T, Fujiki F, Nomura M, Nishida S, Shirakata T, Hosen N, Tsuboi A, Oka Y, Nezu R, Mori M, Doki Y, Aozasa K, Sugiyama H, Oji Y. Over-expression of eukaryotic elongation factor eEF2 in gastrointestinal cancers and its involvement in G2/M progression in the cell cycle. Int. J. Oncol. 2009;34:1181-1189.

[81] Cheng Y, Ren X, Zhang Y, Patel R, Sharma A, Wu H, Robertson GP, Yan L, Rubin E, Yang JM. eEF-2 kinase dictates cross-talk between autophagy and apoptosis induced by Akt Inhibition, thereby modulating cytotoxicity of novel Akt inhibitor MK-2206. Cancer Res. 2011;71:2654-2663.

[82] Lee MH, Surh YJ. eEF1A2 as a putative oncogene. Ann. N.Y. Acad. Sci. 2009;1171:87

[83] Pecorari L, Marin O, Silvestri C, Candini O, Rossi E, Guerzoni C, Cattelani S, Mariani SA, Corradini F, Ferrari-Amorotti G, Cortesi L, Bussolari R, Raschellà G, Federico MR, Calabretta B. Elongation Factor 1 alpha interacts with phospho-Akt in breast cancer cells and regulates their proliferation, survival and motility. Mol. Cancer. Published on line: 3 August 2009. DOI:10.1186/1476-4598-8-58.

[84] Chang R, Wang E. Mouse translation elongation factor eEF1A-2 interacts with Prdx-I to protect cells against apoptotic death induced by oxidative stress. J. Cell. Biochem. 2007;100:267-278.

[85] Amiri A, Noei F, Jeganathan S, Kulkarni G, Pinke DE, Lee JM. eEF1A2 activates Akt and stimulates Akt-dependent actin remodeling, invasion and migration. Oncogene 2007;26:3027-3040.

[86] Tomlinson VA, Newbery HJ, Wray NR, Jackson J, Larionov A, Miller WR, Dixon JM, Abbott CM. Translation elongation factor eEF1A2 is a potential oncoprotein that is over-expressed in two-thirds of breast tumours. BMC Cancer. Published on line: 12 September 2005. DOI:10.1186/1471-2407-5-113.

[87] Graff JR, Konicek BW, Carter JH, Marcusson EG. Targeting the Eukaryotic Translation Initiation Factor 4E for Cancer Therapy. Cancer Res. 2008;68:631-634.

[88] Lazaris-Karatzas A, Smith MR, Frederickson RM, Jaramillo ML, Liu YL, Kung HF, Sonenberg N. Ras mediates translation initiation factor 4E-induced malignant transformation. Genes Dev 1992;6:1631–1642.

[89] Li S, Perlman DM, Peterson MS, Burrichter D, Avdulov S, Polunovsky VA, Bitterman PB. Translation initiation factor 4E blocks endoplasmic reticulum mediated apoptosis. J. Biol. Chem. 2004;279:21312–21317.

[90] Zhang Y, Cheng Y, Zhang L, Ren X, Huber-Keener KJ, Lee S, Yun J, Wang HG, Yang JM. Inhibition of eEF-2 kinase sensitizes human glioma cells to TRAIL and down-regulates Bcl-xL expression. Biochem. Biophys. Res. Commun. 2011;414:129-134.

[91] Caraglia M, Vitale G, Marra M, Budillon A, Tagliaferri P, Abbruzzese A. Alpha-interferon and its effects on signalling pathways within cells. Curr Protein Pept Sci. 2004;5:475-85.

[92] Caraglia M, Passeggio A, Beninati S, Leardi A, Nicolini L, Improta S, Pinto A, Bianco AR, Tagliaferri P, Abbruzzese A. Interferon alpha2 recombinant and epidermal growth factor modulate proliferation and hypusine synthesis in human epidermoid cancer KB cells. Biochem. J. 1997;324:737-741.

[93] Caraglia M, Leardi A, Corradino S, Ciardiello F, Budillon A, Guarrasi R, Bianco AR, Tagliaferri P. alpha-Interferon potentiates epidermal growth factor receptor-mediated effects on human epidermoid carcinoma KB cells. Int J Cancer. 1995;61:342-7.

[94] Lamberti A, Longo O, Marra M, Tagliaferri P, Bismuto E, Fiengo A, Viscomi C, Budillon A, Rapp UR, Wang E, Venuta S, Abbruzzese A, Arcari P, Caraglia M. C-Raf antagonizes apoptosis induced by IFN-alpha in human lung cancer cells by phosphorylation and increase of the intracellular content of elongation factor 1A. Cell Death Differ. 2007;14:952-962.

[95] Bunai F, Ando K, Ueno H, Numata O. () Tetrahymena eukaryotic translation elongation factor 1A (eEF1A) bundles filamentous actin through dimer formation. J Biochem. 2006;140:393-399.

[96] Sanges C, Scheuermann C, Zahedi RP, Sickmann A, Lamberti A, Migliaccio N, Baljuls A, Marra M, Zappavigna S, Rapp U, Abbruzzese A, Caraglia M, Arcari P. Raf kinases mediate the phosphorylation of eukaryotic translation elongation factor 1A and regulate its stability in eukaryotic cells. Cell Death Dis. Published online 1 March 2012. DOI:10.1038/cddis.2012.16.

Chapter 5

Apoptosis During Cellular Pattern Formation

Masahiko Takemura and Takashi Adachi-Yamada

Additional information is available at the end of the chapter

1. Introduction

We can all see a variety of ordered cellular patterns consisting of various cell types, throughout nature. It is surprising that these ordered cellular patterns are created reproducibly during development in all individuals. Elucidating their underlying molecular mechanisms has been an interesting research subject for developmental biologists. The essential building blocks in these processes are cell proliferation, cell shape change, cell movement, and apoptosis. These cellular behaviors must be coordinated through cell-cell communication.

Drosophila melanogaster has provided many insights into the underlying mechanisms of many biological processes, including tissue patterning. One typical example is the ordered pattern of bristles which cover the whole body of the adult fly. In these bristles, the thorax and wing margin ones (Figure 1A, 1C, and 1D) have been examined extensively for this purpose. Specification and development of these bristles have been well studied [1]. When a single sensory organ precursor (SOP) is specified in an epithelial field, the SOP prevents its neighboring cells from choosing the same cell fate by activating Notch signaling there. This Notch-mediated lateral inhibition ensures the proper number and spatial separation of SOPs. The SOP then undergoes a series of asymmetric cell divisions, producing the components of sensory bristles, such as a shaft, socket, sheath, glial cell and neuron. However, whether the lateral inhibition is sufficient to create the final intricate pattern of bristle distribution is unknown.

Apoptosis is used extensively to refine developing structures, such as in formation of vertebrate digits and sculpting of the insect wing [2]. Apoptosis also contributes to tissue patterning by removing abnormal cells [3-5] and eliminating excess populations of cells [6].

Difference in adhesiveness between cell types is another important factor in tissue patterning. Differential adhesion mediated by heterophilic adhesion molecules forces cells to rearrange during development. For example, in the oviduct epithelium of the Japanese quail, two distinct types of columnar cells; goblet-type gland cells and ciliated cells are arranged in a checkerboard pattern (Figure 1E) [7]. Preferential adhesion between different cell types rather than between cells of the same type could account for this pattern [8].

Experiments have shown that spatial and temporal regulation of apoptosis or cell adhesion is indispensable for correct patterning. Inappropriate cells must be removed at the proper time by apoptosis and each living cell must attach properly to its counterparts. How are these processes regulated? In this chapter, we will describe the *Drosophila* eye and posterior wing margin, which are interesting tissues showing geometrically ordered repetitive cellular arrangements (Figure 1B, 1C, and 1D). We first describe their unique cellular arrangement and then follow the patterning process. We will also explain the underlying mechanisms, which seem to be conserved in both tissues. Furthermore, we will end with a discussion of the cellular patterning of the mammalian organ of Corti.

2. *Drosophila* eye patterning

A striking example of ordered cellular packing is the *Drosophila* compound eye. It is comprised of approximately 750–800 ommatidia, which are arranged in a hexagonal close packing manner. Each ommatidium contains eight photoreceptors, four cone cells, and two primary pigment cells. At the early stage, each ommatidium is surrounded by a few layers of the interommatidial precursor cells (IPCs). These cells undergo dynamic cell rearrangement and eventually differentiate into the secondary and tertiary pigment cells which optically insulate ommatidia, and the mechanosensory bristles (Figure 2). Apoptosis plays an important role in this cell rearrangement [10]. Approximately one-third of IPCs are eliminated through apoptosis between 24 and 40 hours after puparium formation (APF) [11, 12]. As a result of apoptosis, only a single layer of IPCs surrounds each ommatidium.

The remaining two-thirds of the interommatidial cells, which are in contact with ommatidia, do not undergo apoptosis. Spatial regulation of apoptosis is mediated by epidermal growth factor receptor (EGFR) signaling [13]. Spitz, a ligand of EGFR, is produced in the primary pigment cells and secreted around surrounding cells. This activation of EGFR signaling downregulates the activity of Hid, a proapoptotic protein, which prevents these adjacent cells from undergoing apoptosis [14, 15]. In this fashion, only non-adjacent IPCs lack the EGFR signal and thus undergo apoptosis.

3. *Drosophila* wing margin hairs

The *Drosophila* posterior wing margin hair is another interesting example of ordered arrangement of cells. They are aligned along the posterior wing margin with two rows in a zigzag manner. Recently, we elucidated the patterning process of the posterior wing margin hairs [16]. Since thsese wing margin hairs are comprised of only shaft and socket cells [17], they do not work as sensors; rather they may affect airflow over the surface of the wing or protect the wing margin. We call both shaft and socket cells "hair cells" here. The zigzag alignment patterning of the hairs also requires apoptosis-related cell rearrangement as seen in the eye patterning described above. At an early stage of pupal development (20 hours

APF), hair cells are not positioned in a zigzag manner, but rather at random. However, the rearrangement of wing margin cells occurs and, by 30 hours APF, the zigzag pattern is created (Figure 3). After apoptosis, wing margin cells can be classified into two distinct types: one is 'interhair cell' which is aligned in the same row of hair cells alternatively, and 'tooth cell' which is located on the dorsoventral boundary side of hair cells and named after the teeth of a zipper. The zigzag pattern of hair cells is ensured by interlocking arrangement of these two kinds of cells. As a result of the cell rearrangement, the dorsal and ventral edges interlock.

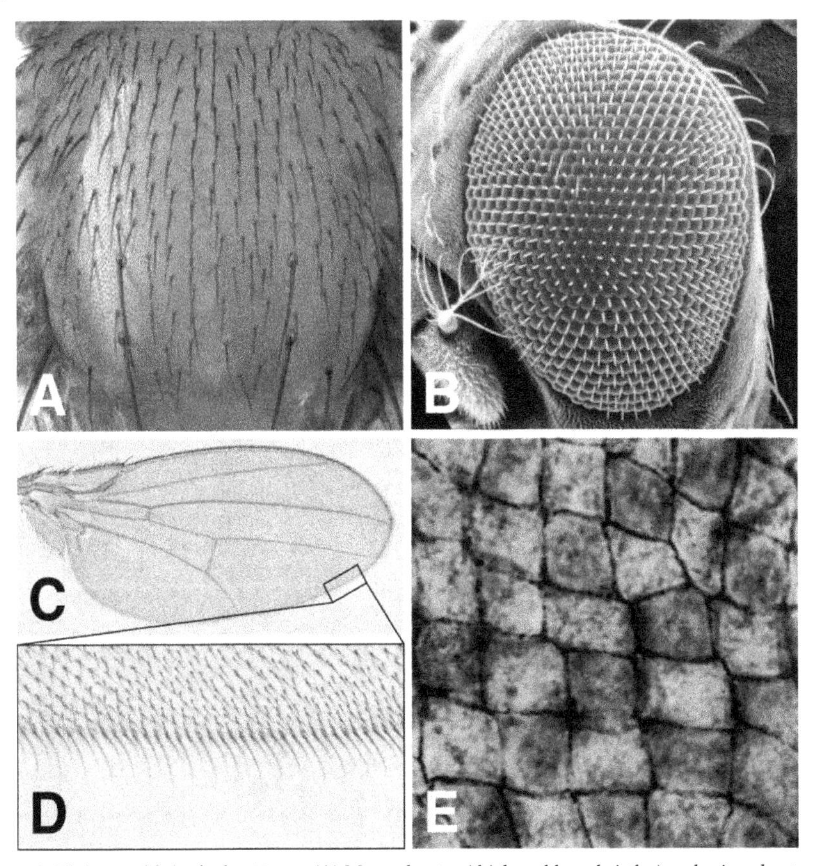

Figure 1. Elaborate biological patterns. (**A**) Macrochaetae (thick and long bristles) and microchaetae (thin and short bristles) on the *Drosophila* adult notum shows a stereotypical pattern. (**B**) SEM image of the *Drosophila* compound eye, which is comprised of 750–800 ommatidia arranged in a precise honeycomb-like pattern. Mechanosensory bristles are found at alternate vertices of the hexagonal array. (**C**) *Drosophila* adult wing. (**D**) Higher magnification view of the posterior wing margin (the boxed region in C). Wing margin hairs are aligned in even intervals. (**E**) Oviduct epithelium of the Japanese quail shows a checkerboard pattern, comprised of goblet-type gland cells and ciliated cells. This image is adapted from Honda *et al.* [9].

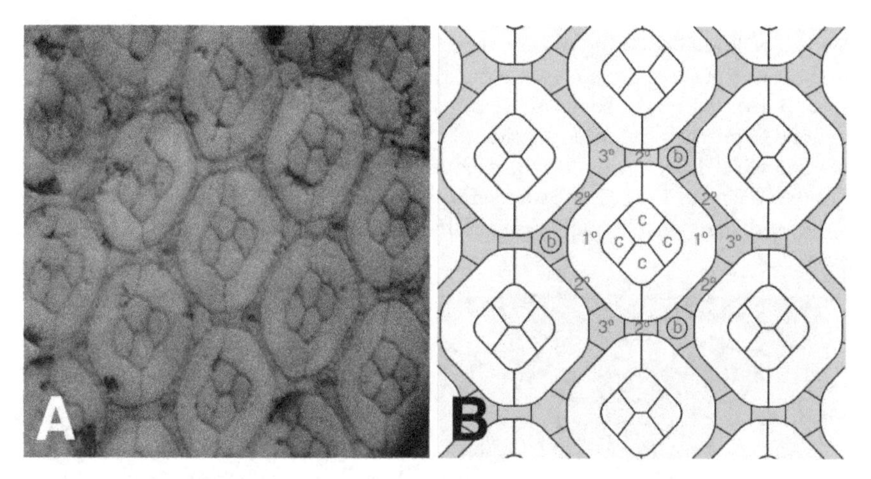

Figure 2. ***Drosophila*** **pupal retina.** (**A**) Cobalt sulfide staining of pupal eye. The photoreceptor cells are out of focus in this picture. (**B**) Schematic drawing of pupal eye. Each ommatidium is surrounded by the secondary and tertiary pigment cells, and bristles. Abbreviations: c, cone cell; 1°, primary pigment cell; 2°, secondary pigment cell; 3°, tertiary pigment cell; b, bristle.

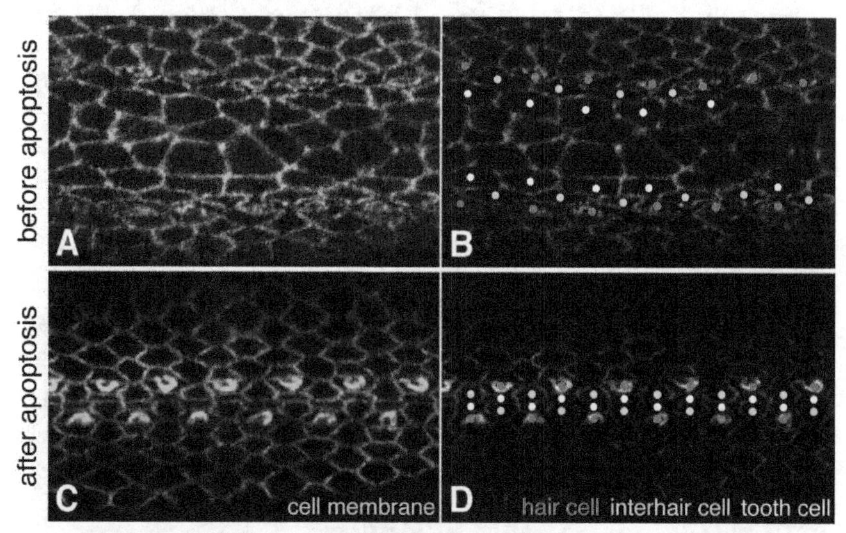

Figure 3. Zigzag pattern formation of wing margin hairs. Cellular arrangement of the *Drosophila* posterior wing margin before (A, B: 20 hours APF) and after (C, D: 30 hours APF) rearrangement. Cell membrane is marked by E-cadherin expression (green in A, C; white in B, D). Hair cells (shaft and socket), interhair cells, and tooth cells are marked by magenta, cyan, and yellow circles, respectively (B, D). (**A, B**) Before apoptosis, posterior wing margin cells are not arranged in an ordered manner. At this stage, hair cells are not positioned in a zigzag manner. (**C, D**) After cell rearrangement, hair cells, interhair cells, and tooth cells establish their unique cell shapes and are aligned in a surprisingly ordered manner. Note that the double row of hair cells is aligned in a zigzag manner.

During this cell rearrangement, a subset of wing margin cells is removed through apoptosis. The dying cells are the cells that have not attached to the hair cells. Blocking apoptosis by expressing the baculovirus caspase inhibitor p35 [18] in wing margin cells using the GAL4/UAS system [19] inhibits cell rearrangement, indicating that apoptosis is required for this process.

What triggers apoptosis in a precise temporal manner? Ecdysone, an insect steroid hormone, is indispensable for progression in most of the developmental stages of *Drosophila*. In addition, ecdysone triggers several apoptotic events associated with metamorphosis through binding with ecdysone receptors and induces expression of proapoptotic genes, such as *hid* and *reaper* [20]. During insect metamorphosis, many larval tissues, including the salivary gland and midgut, are eliminated through apoptosis. Blocking ecdysone signaling by expressing a dominant-negative form of the ecdysone receptor in wing margin cells results in excess wing margin cells, indicating that apoptosis is blocked. Thus, ecdysone signaling is required for inducing apoptosis in wing margin cells as well as in other tissues during metamorphosis.

Then, how is apoptosis in the developing tissue regulated in a precise spatial manner? Vein, a diffusible ligand of EGFR, is expressed specifically in the hair cells. In addition, EGFR activation is observed in cells surrounding hair cells, as revealed by the expression of *sprouty*, which is a target gene of EGFR signaling in the wing. These results indicate that EGFR signaling is activated in a paracrine manner in the posterior wing margin and that its activation pattern correlates with the cell survival pattern there. To confirm the relationship between the activation of EGFR signaling and cell survival of wing margin cells, we used the TARGET system, which combines the GAL4/UAS system [19] and a temperature-sensitive version of GAL80, a GAL4 inhibitor [21]. This system allows us to induce transgene expression in a desired spatiotemporal manner. Activation of EGFR signaling by expressing Ras^{V12}, a constitutively active form of its downstream activator Ras [22], in the wing margin during 0~30 hours APF, results in excess wing margin cells. On the other hand, knockdown of EGFR by expressing an inverted repeat complementary to *Egfr* mRNA (for RNAi) or a dominant negative form of EGFR results in ectopic apoptosis. Both of these genetic manipulations result in disruption of the zigzag pattern. Taken together, we found that the hair cell produces the signaling ligand molecule Vein for cell survival, which allows the surrounding cells that receive the ligands to survive.

4. Coordination of preferential adhesion and secreted survival signaling molecules

In both tissues described above, locally diffusible ligands are used to make neighboring cells survive. Thus, regulation of cell-cell contact is an important factor for controlling the spatial pattern of apoptosis. In both cases, *Drosophila* NEPH1/Nephrin homologs, which are transmembrane proteins belonging to the immunoglobulin superfamily, are involved in the preferential adhesion between cell types. In humans, mutations in the nephrin gene are associated with the congenital nephrotic syndrome of the Finnish type [23]. The

NEPH1/Nephrin homologs can be classified into two subfamilies, NEPH1 and Nephrin, that regulate many biological processes through heterophilic cell adhesion between NEPH1 and Nephrin groups, including myoblast fusion [24, 25], axonal pathfinding in the visual system [26-28], retinotopic map formation [29]. The NEPH1/Nephrin homologs are also involved in the formation of a slit diaphragm-like structure in the *Drosophila* nephrocyte, an analog of the mammalian podocyte in the kidney [30, 31]. In *Drosophila,* there are four members of the NEPH1/Nephrin subfamilies: two members of the NEPH1 subfamily, *kin of irreC* (*kirre,* also known as *dumbfounded*) and *roughest* (*rst*), and two members of the Nephrin subfamily, *hibris* (*hbs*) and *sticks-and-stones* (*sns*) [32-34, 27, 35, 36].

In the compound eye, immunohistochemical staining reveals that all four molecules accumulate at the interface between ommatidia and IPCs [37, 38]. Hibris and Sns are expressed in the ommatidia, and their binding partners Kirre and Rst are expressed in the IPCs. Computer simulation have shown that preferential adhesion between ommatidia and IPCs contribute to the cell rearrangement [10].

Similarly, antibody staining of the pupal wing indicate that all these adhesion molecules accumulate the interface between hair cell and their neighboring cells [16]. Enhancer-trap reporters for these genes also show that NEPH1 groups and Nephrin homologs are expressed in an almost complementary pattern (Figure 4). Cell-type-specific knockdown of these molecules results in disruption of the wing margin pattern. For instance, when we knockdown Rst in interhair cells, we observed some hair cells away from interhair cells and tooth cells (unpublished data). Knockdown of each gene results in disruption of the posterior wing margin hairs, indicating that all four of these molecules are required for proper hair patterning.

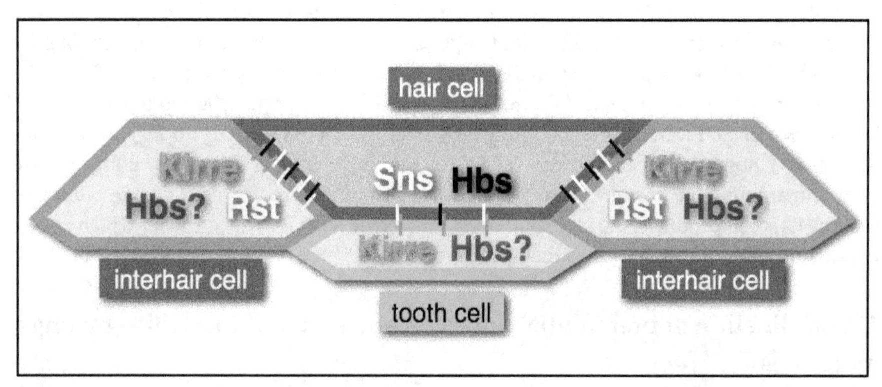

Figure 4. Expression pattern of NEPH1/Nephrin homologs in the *Drosophila* posterior wing margin. Schematic drawing of the allocation of hair cell and the neighboring wing margin cells (interhair/tooth cells). Nephrin homologs (Sns and Hbs) are expressed in the hair cell and NEPH1 homologs (Kirre and Rst) are expressed in the interhair cell and tooth cell. These molecules accumulate at the interface between hair cells and interhair cells or tooth cells. This complementary expression pattern of these heterophilic adhesion molecules contributes to the attachment between hair cells and interhair/tooth cells.

Therefore, in both the *Drosophila* eye and posterior wing margin, apoptosis-dependent cell rearrangement is strictly regulated by secretion of EGFs and preferential adhesion between cell types through heterophilic adhesion molecules.

This seems to be a good strategy for creating ordered repetitive cellular patterns through refinement (Figure 5). It is tempting to speculate that similar mechanisms work in other tissues of other organisms. Lastly, we will discuss the cellular patterning in the cochlea, a mammalian inner ear organ.

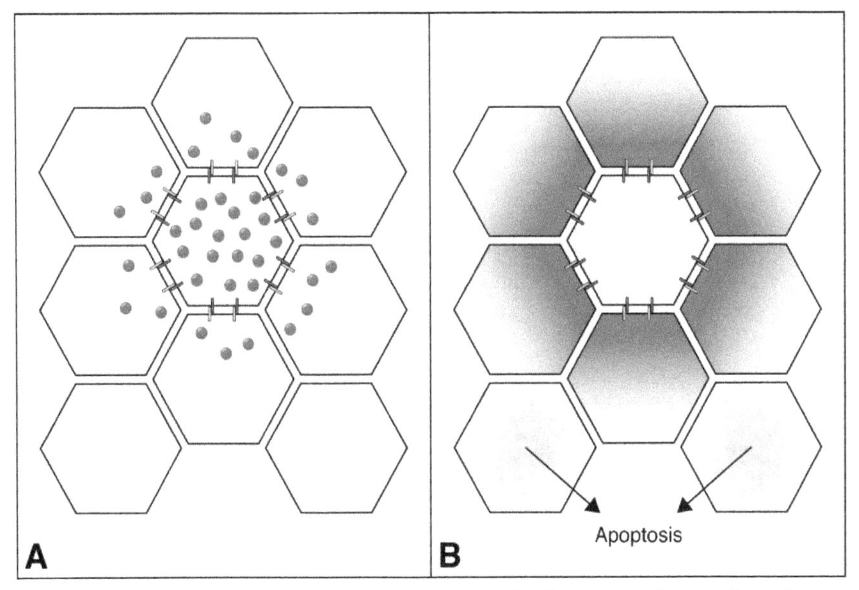

Figure 5. Model for cooperation between preferential cell adhesion and locally-secreting cell survival signal. (A) Locally-diffusing signaling molecules (magenta circles) secreted by a cell (yellow). Heterophilic adhesion molecules (light or dark green bars) attach the cell to its neighboring cells. (B) Locally-diffusing signaling molecules required for cell survival allow cells that neighbor the signaling center to survive. Cells that do not receive enough cell survival signaling molecules are destined to undergo apoptosis (cyan cells).

5. Patterning of outer hair cells and supporting cells in the organ of Corti

The sensory epithelium of the mammalian cochlea, the organ of Corti, has three rows of outer hair cells and supporting cells, which are aligned in a checkerboard pattern (Figure 6). Outer hair cells (OHCs) are essential for the amplification of sound [39] and the loss of these cells can lead to hearing loss. Specification of hair cells and supporting cells are mediated by Notch-mediated lateral inhibition [40-42], as is the case with *Drosophila* bristles. Recently, another adhesion molecules of the immunoglobulin superfamily, nectins, were found to be involved in patterning of the inner ear [43]. Mammals have four distinct nectins that mediate both homophilic adhesion and heterophilic adhesion with nonidentical nectins.

Heterophilic adhesion is stronger than homophilic adhesion. Nectin-1 is expressed in outer hair cells, nectin-3 in supporting cells, and nectin-2 in both. In the nectin-1 or nectin-3 KO mice, disruption of the checkerboard pattern is observed. Although the contribution of apoptosis for this cellular pattern remains unknown, it is a reasonable hypothesis that apoptosis may contribute to removal of excess OHCs and supporting cells. Thus, similar refining mechanisms of cellular arrangement as describe above may be conserved across species in a wide variety of multicellular organisms.

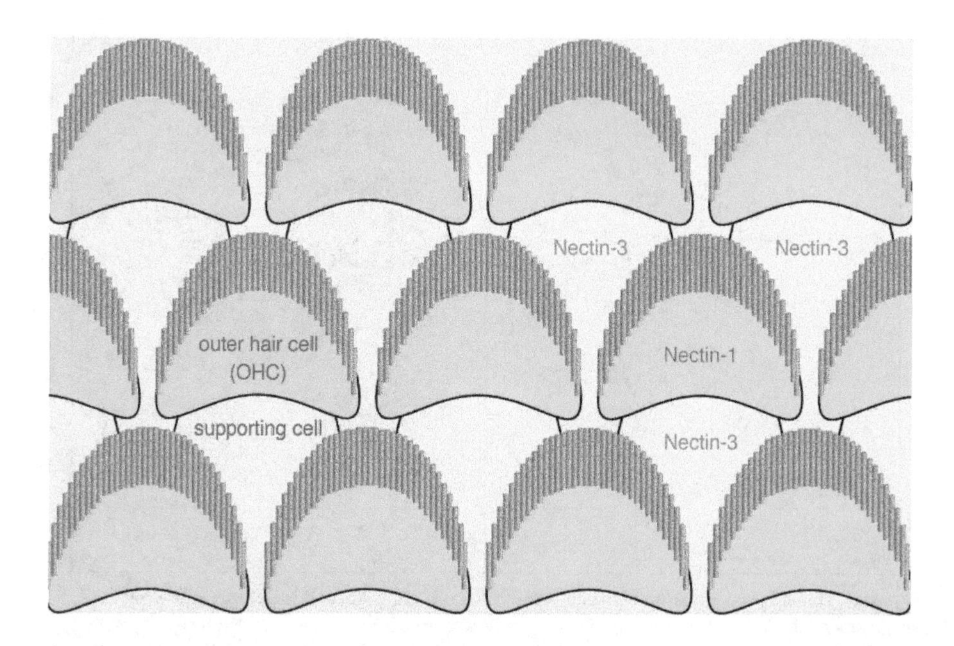

Figure 6. Schematic drawing of the arrangement of OHCs and supporting cells in the organ of Corti. Three rows of OHCs and supporting cells are arranged in an alternatively distributed pattern. This pattern can be formed by heterophilic adhesion between Nectini-1 and Nectin-3, which are expressed in OHC and supporting cell, respectively.

6. Conclusion

We have described the apoptosis-dependent cell rearrangement for refining cellular arrangements of the *Drosophila* eye and posterior wing margin. In both tissues, secreted extracellular signaling molecules to promote cell survival and heterophilic adhesion molecules are involved in correct attachment between the various cell types, which are essential for patterning. This strategy seems to reasonably achieve the geometrically ordered packing of cells and may be a conserved method in other tissues of other organisms, such as the mammalian organ of Corti.

Author details

Masahiko Takemura
Department of Genetics, Cell Biology and Development, University of Minnesota, Minneapolis, Minnesota, USA

Department of Life Science, Faculty of Science, Gakushuin University, 1-5-1 Mejiro, Toshima-ku, Tokyo, Japan

Takashi Adachi-Yamada *
Department of Life Science, Faculty of Science, Gakushuin University, 1-5-1 Mejiro, Toshima-ku, Tokyo, Japan

Acknowledgement

We would like to thank Dr. Hisao Honda for kindly providing the picture of the oviduct epithelium of the Japanese quail. We are grateful to Daniel Levings and Pui Choi for critical reading of the manuscript.

7. References

[1] Lai EC, Orgogozo V (2004) A hidden program in Drosophila peripheral neurogenesis revealed: fundamental principles underlying sensory organ diversity. Dev Biol 269(1):1–17
[2] Baehrecke EH (2002) How death shapes life during development. Nat Rev Mol Cell Biol 3(10):779–787
[3] Adachi-Yamada T, O'Connor MB (2002) Morphogenetic apoptosis: a mechanism for correcting discontinuities in morphogen gradients. Dev Biol 251(1):74–90
[4] Takemura M, Adachi-Yamada T (2011) Repair responses to abnormalities in morphogen activity gradient. Development, Growth & Differentiation 53(2):161–167
[5] Adachi-Yamada T (2004) Mechanisms for Removal of Developmentally Abnormal Cells: Cell Competition and Morphogenetic Apoptosis. Journal of Biochemistry 136(1):13–17
[6] Koto A, Kuranaga E, Miura M (2011) Apoptosis Ensures Spacing Pattern Formation of Drosophila Sensory Organs. Curr Biol 21(4):278–287
[7] Yamanaka H (1990) Pattern formation in the epithelium of the oviduct of Japanese quail. Int J Dev Biol 34(3):385–390
[8] Yamanaka HI, Honda H (1990) A checkerboard pattern manifested by the oviduct epithelium of the Japanese quail. Int J Dev Biol 34(3):377–383
[9] Honda H, Yamanaka H, Eguchi G (1986) Transformation of a polygonal cellular pattern during sexual maturation of the avian oviduct epithelium: computer simulation. J Embryol Exp Morphol 98:1–19

* Corresponding Author

[10] Larson DE, Johnson RI, Swat M, Cordero JB, Glazier JA, Cagan RL (2010) Computer Simulation of Cellular Patterning Within the Drosophila Pupal Eye. PLoS Comput Biol 6(7):e1000841

[11] Cagan RL, Ready DF (1989) The emergence of order in the Drosophila pupal retina. Dev Biol 136(2):346–362

[12] WOLFF T, Ready DF (1991) Cell death in normal and rough eye mutants of Drosophila. Development 113(3):825–839

[13] Brachmann CB, Cagan RL (2003) Patterning the fly eye: the role of apoptosis. Trends Genet 19(2):91–96

[14] Kurada P, White K (1998) Ras promotes cell survival in Drosophila by downregulating hid expression. Cell 95(3):319–329

[15] Bergmann A, Agapite J, McCall K, Steller H (1998) The Drosophila gene hid is a direct molecular target of Ras-dependent survival signaling. Cell 95(3):331–341

[16] Takemura M, Adachi-Yamada T (2011) Cell death and selective adhesion reorganize the dorsoventral boundary for zigzag patterning of Drosophila wing margin hairs. Dev Biol 357(2):336–346

[17] Hartenstein V, Posakony JW (1989) Development of adult sensilla on the wing and notum of Drosophila melanogaster. Development 107(2):389–405

[18] HAY B, WOLFF T, RUBIN G (1994) Expression of baculovirus P35 prevents cell death in Drosophila. Development 120(8):2121–2129

[19] Brand AH, Perrimon N (1993) Targeted gene expression as a means of altering cell fates and generating dominant phenotypes. Development 118(2):401–415

[20] Yin VP, Thummel CS (2005) Mechanisms of steroid-triggered programmed cell death in Drosophila. Semin Cell Dev Biol 16(2):237–243

[21] McGuire SE, Le PT, Osborn AJ, Matsumoto K, Davis RL (2003) Spatiotemporal rescue of memory dysfunction in Drosophila. Science 302(5651):1765–1768

[22] Karim FD, Rubin GM (1998) Ectopic expression of activated Ras1 induces hyperplastic growth and increased cell death in Drosophila imaginal tissues. Development 125(1):1–9

[23] Kestilä M, Lenkkeri U, Männikkö M, et al (1998) Positionally cloned gene for a novel glomerular protein--nephrin--is mutated in congenital nephrotic syndrome. Mol Cell 1(4):575–582

[24] Galletta BJ, Chakravarti M, Banerjee R, Abmayr SM (2004) SNS: Adhesive properties, localization requirements and ectodomain dependence in S2 cells and embryonic myoblasts. Mech Dev 121(12):1455–1468

[25] Shelton C, Kocherlakota KS, Zhuang S, Abmayr SM (2009) The immunoglobulin superfamily member Hbs functions redundantly with Sns in interactions between founder and fusion-competent myoblasts. Development 136(7):1159–1168

[26] Fischbach K-F, Linneweber GA, Felix Malte Andlauer T, Hertenstein A, Bonengel B, Chaudhary K (2009) The Irre Cell Recognition Module (IRM) Proteins. J Neurogenet 23(1-2):48–67

[27] Ramos RG, Igloi GL, Lichte B, Baumann U, Maier D, Schneider T, Brandstätter JH, Fröhlich A, Fischbach KF (1993) The irregular chiasm C-roughest locus of Drosophila,

which affects axonal projections and programmed cell death, encodes a novel immunoglobulin-like protein. Genes Dev 7(12B):2533–2547
[28] Schneider T, Reiter C, Eule E, Bader B, Lichte B, Nie Z, Schimansky T, Ramos RG, Fischbach K-F (1995) Restricted expression of the irreC-rst protein is required for normal axonal projections of columnar visual neurons. Neuron 15(2):259–271
[29] Sugie A, Umetsu D, Yasugi T, Fischbach K-F, Tabata T (2010) Recognition of pre- and postsynaptic neurons via nephrin/NEPH1 homologs is a basis for the formation of the Drosophila retinotopic map. Development 137(19):3303–3313
[30] Zhuang S, Shao H, Guo F, Trimble R, Pearce E, Abmayr SM (2009) Sns and Kirre, the Drosophila orthologs of Nephrin and Neph1, direct adhesion, fusion and formation of a slit diaphragm-like structure in insect nephrocytes. Development 136(14):2335–2344
[31] Weavers H, Prieto-Sánchez S, Grawe F, Garcia-López A, Artero R, Wilsch-Bräuninger M, Ruiz-Gómez M, Skaer H, Denholm B (2009) The insect nephrocyte is a podocyte-like cell with a filtration slit diaphragm. Nature 457(7227):322–326
[32] Artero RD, Castanon I, Baylies MK (2001) The immunoglobulin-like protein Hibris functions as a dose-dependent regulator of myoblast fusion and is differentially controlled by Ras and Notch signaling. Development 128(21):4251–4264
[33] Bour BA, Chakravarti M, West JM, Abmayr SM (2000) Drosophila SNS, a member of the immunoglobulin superfamily that is essential for myoblast fusion. Genes Dev 14(12):1498–1511
[34] Dworak HA, Charles MA, Pellerano LB, Sink H (2001) Characterization of Drosophila hibris, a gene related to human nephrin. Development 128(21):4265–4276
[35] Ruiz-Gómez M, Coutts N, Price A, Taylor MV, Bate M (2000) Drosophila dumbfounded: a myoblast attractant essential for fusion. Cell 102(2):189–198
[36] Strünkelnberg M, Bonengel B, Moda LM, Hertenstein A, de Couet HG, Ramos RG, Fischbach KF (2001) rst and its paralogue kirre act redundantly during embryonic muscle development in Drosophila. Development 128(21):4229–4239
[37] Bao S, Cagan RL (2005) Preferential adhesion mediated by Hibris and Roughest regulates morphogenesis and patterning in the Drosophila eye. Dev Cell 8(6):925–935
[38] Bao S, Fischbach K-F, Corbin V, Cagan RL (2010) Preferential adhesion maintains separation of ommatidia in the Drosophila eye. Dev Biol 344(2):948–956
[39] Dallos P, Wu X, Cheatham MA, et al (2008) Prestin-Based Outer Hair Cell Motility Is Necessary for Mammalian Cochlear Amplification. Neuron 58(3):333–339
[40] Eddison M, Le Roux I, Lewis J (2000) Notch signaling in the development of the inner ear: lessons from Drosophila. Proc Natl Acad Sci USA 97(22):11692–11699
[41] Müller U, Littlewood-Evans A (2001) Mechanisms that regulate mechanosensory hair cell differentiation. Trends Cell Biol 11(8):334–342
[42] Lanford PJ, Lan Y, Jiang R, Lindsell C, Weinmaster G, Gridley T, Kelley MW (1999) Notch signalling pathway mediates hair cell development in mammalian cochlea. Nat Genet 21(3):289–292

[43] Togashi H, Kominami K, Waseda M, Komura H, Miyoshi J, Takeichi M, Takai Y (2011) Nectins Establish a Checkerboard-Like Cellular Pattern in the Auditory Epithelium. Science 333(6046):1144–1147

Chapter 6

Extra-Telomeric Effects of Telomerase (hTERT) in Cell Death

Gregory Lucien Bellot and Xueying Wang

Additional information is available at the end of the chapter

1. Introduction

1.1. Telomeres

Telomeres are heterochromatic structures found at the ends of chromosomes which are involved in the protection of chromosomes from degradation and DNA-repair mechanisms (Moyzis et al., 1988; Shay and Wright, 2004; Wyatt et al., 2010). Discovery of telomeres also solved the "end-replication problem" which was exposed after the observation that the 3′-extremity of chromosomes was not completely replicated during each cell cycle. As a consequence telomeres play a fundamental role in chromosomes and the overall genome stability (de Lange, 2005; Martinez and Blasco, 2011; O'Sullivan and Karlseder, 2010; Takai et al., 2003). In mammals telomeres are composed of tandem repeats of the oligonucleotide sequence TTAGGG and bound by a composite structure of proteins named the shelterin complex (de Lange, 2005, 2010; Diotti and Loayza, 2011; Longhese et al., 2012; Martinez and Blasco, 2010; O'Sullivan and Karlseder, 2010). In somatic cells telomeres are shortened after each division cycle and when a critical short length has been reached then the cell replication stops before these cells undergo senescence or apoptosis (Counter, 1996; Deng and Chang, 2007). This mechanism is supposed to be responsible for the "Hayflick limit" which corresponds to the number of times a cell can divide before it stops proliferating (Deng and Chang, 2007; Hayflick, 1965; Hayflick and Moorhead, 1961). The phenomenon of telomeres shortening is directly linked to the ageing process by acting as a mitotic clock and by inducing senescence and/or apoptosis once the Hayflick limit has been reached (Blasco, 2003; Djojosubroto et al., 2003; Goronzy et al., 2006; Liew et al., 2009; Martinez and Blasco, 2010; Shin et al., 2006).

1.2. Telomerase

The main mechanism involved in telomere maintenance and *de novo* synthesis of telomeric DNA is represented by the activity of the telomerase holoenzyme. Telomerase is responsible for the addition of the telomere repeats TTAGGG at the end of chromosomes (Blackburn et al., 1989; Greider and Blackburn, 1985, 1987). The catalytic core of telomerase is a ribonucleoprotein consisting of a reverse transcriptase (TERT) and a telomerase RNA template (TERC) whereas other species-specific co-factors may be required to form the whole holoenzyme (Martinez and Blasco, 2011; Wyatt et al., 2010). The catalytic subunit TERT is comprised of 3 main domains. The N-terminal extremity contains two domains called the telomerase essential N-terminal domain (TEN) and the telomerase RNA-binding domain (TRBD) which are involved in the association of TERC with TERT. The central part of the protein contains the catalytic domain for reverse transcriptase activity (RT) with seven conserved motifs which are essential for the enzymatic activity. The sequences of both N-terminal extremity and the core catalytic domain of TERT are evolutionarily conserved among species. On the other hand, the C-terminal domain displays a higher variability and therefore may be related to species-specific function (Wyatt et al., 2010).

While it appears that the primary function of telomerase is to elongate telomeres by adding telomeric DNA at the end of chromosomes, many studies in the past decade has started to uncover other potentially crucial functions of telomerase besides its direct role in telomeres maintenance (De Semir et al., 2007; Gordon and Santos, 2010; Majerska et al., 2011; Martinez and Blasco, 2011). As a matter of fact it has been observed that telomerase is able to promote tumor oncogenic transformation independently of its ability to elongate telomeres (Stewart et al., 2002) and appears to be involved in the modulation of mechanisms related to cell survival, genes regulation, cell signaling, cell proliferation and differentiation, metabolism, and DNA repair (De Semir et al., 2007; Lai et al., 2007; Majerska et al., 2011; Saretzki, 2009). Such functions have been described as extra-telomeric roles of telomerase. While these non-telomeric functions still remain generally enigmatic, their relationship with the regulation of crucial cellular mechanisms emphasize the critical importance of investigating this field in order to improve our understanding of telomerase biology. This chapter proposes to highlight and summarize the current knowledge about the non-telomeric effects of the catalytic subunit TERT and more particularly its related roles to cell death regulation, with regard to the relationships between TERT and key actors of apoptosis, i.e. mitochondria, oxidative stress and p53.

2. Telomeres dysfunctions and relationship to diseases

Telomeres alterations have been described in many diseases including aging-related disease (Hiyama and Hiyama, 2007; Martinez and Blasco, 2011; O'Sullivan and Karlseder, 2010) (Figure 1). Some inherited diseases such as Dyskeratosis Congenital and aplastic anaemia results in impaired telomerase activity leading to major bone marrow failures and premature ageing syndromes (Calado et al., 2002; Calado et al., 2009; Mason et al., 2005; Mitchell et al., 1999; Shtessel and Ahmed, 2011; Vulliamy et al., 2002). More recently it was observed that mutations in telomerase components TERT or TERC may be linked in the occurrence of the idiopathic pulmonary fibrosis resulting in dramatic destruction of lung tissues (Alder et al., 2008; Tsakiri et al., 2007). Other studies pointed the direct correlation between telomere dysfunction and pathologies such as cardiovascular disease, carotid atherosclerosis and increased insulin resistance (Benetos et al., 2004; Epel et al., 2006; Gardner et al., 2005; Kuhlow et al., 2010). In addition it has been demonstrated that telomere dysfunction in chronically stressed patients may lead to premature immune ageing (Damjanovic et al., 2007; Goronzy et al., 2006) (Figure 1).

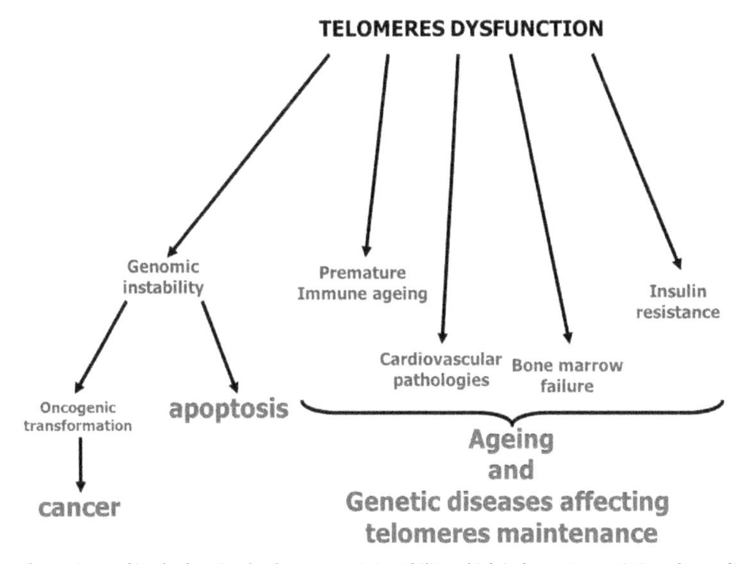

Telomeres shortening and/or dysfunction leads to genomic instability which induces apoptosis in order to eliminate such aberrant cells. However genomic instability can promote oncogenic transformation and lead to the appearance of a cancerous phenotype. Telomeres dysfunction seen in genomic diseases affecting telomeres maintenance have been shown to induce pathologies related to premature aging thus resulting in deficient immune system, bone marrow failure or rise in insulin resistance.

Figure 1. Relationship between telomeres dysfunction and pathologies.

While differentiated tissues display relatively low telomerase activity (Wright et al., 1996), it has been widely observed that malignant cells from a large variety of cancers present a significantly increased telomerase activity. Around 90% of tumors have been reported to be telomerase-positive tumors, thus making telomerase the most widely expressed gene across all types of cancer (Shay and Bacchetti, 1997). It appears that telomerase is a major protein that holds the key to infinite proliferative capacity which is a necessary step toward oncogenic transformation that has been described as one of the hallmarks of cancer (Hanahan and Weinberg, 2011).

The high telomerase activity levels in cancer correlate directly with malignant and metastatic potential (Oishi et al., 1998; Pirker et al., 2003). As a consequence, telomerase has become a promising target in the race to the development of new anti-cancer therapies. Therefore, it is of critical importance to understand the roles of telomerase and telomeres in cancer development in order to design these new anti-cancer strategies. Some telomerase-based approaches have been developed in the recent years such as gene therapy, immunotherapy and small-molecule inhibitors of telomerase (Keith et al., 2007; Shay and Keith, 2008). Some of these promising candidates for telomerase-based therapies are now in different phases of clinical trials (Harley, 2008; Ouellette et al., 2011; Shay and Wright, 2011).

The understanding of the regulation of telomerase appears to be an important issue that may help to improve therapies related to pathologies mentioned above from inherited diseases to ageing-related diseases and cancers. In order to improve these telomerase-based approaches there is a crucial need to investigate closely the functions of telomerase as a mean to understand the full extent of the roles in which telomerase is involved.

3. Extra-telomeric functions of TERT and its implication in cell death

3.1. TERT, oxidative stress and mitochondria

Mitochondria are key organelles of the cell as it is a major metabolic centre and mitochondrial dysfunctions are linked to many pathologic syndromes. Mitochondria hold a primary role in cell biology through its implication in energetic metabolism, production of reactive oxygen species (ROS) and also as a key regulator of apoptosis (Fogg et al., 2011; Low et al., 2011; Saretzki, 2009). The mitochondrial pathway of apoptosis, also known as the intrinsic pathway, leads to the release of apoptogenic proteins from the intermembrane space of mitochondria upon apoptotic stimuli which in turn results in the activation of caspase 9 through the formation of a protein complex called the apoptosome (Antonsson, 2004; Saelens et al., 2004; Yuan et al., 2011). This mechanism is regulated by a family of proteins called the Bcl-2 family of proteins, which are responsible for the regulation of the apoptotic mitochondrial pathway through the activation of caspases (Antonsson, 2004). On the other hand it is also of critical importance to understand that mitochondria are a major producer of ROS which activate many downstream pathways involved in the modulation of mechanisms such as cell death or cell survival, cell proliferation, senescence and ageing (Indran et al., 2011; Saretzki, 2009).

Interestingly in the past decade the initial relationship between increased of oxidative stress, telomeres shortening and ageing leads to the investigation of a potential connection between telomerase and mitochondria (Saretzki, 2009; Saretzki et al., 2003). Following this hypothesis it was then demonstrated that telomerase, or more specifically the catalytic subunit TERT is able to translocate from the nucleus to the mitochondria following drug treatments or increase of oxidative stress (Ahmed et al., 2008; Haendeler et al., 2009; Santos et al., 2006; Saretzki, 2009) (Figure 2). This new interesting finding was linked to the discovery of mitochondrial targeting sequence at the N-terminal extremity of TERT (Santos et al., 2004). It appears then that TERT localization is a dynamic and regulated mechanism which is induced as a response to environmental stress. It has been shown that oxidative stress can drive 80% to 90% of endogenous TERT to mitochondria and that this phenomenon does not involve *de novo* synthesis of TERT (Ahmed et al., 2008). However this function of mitochondrial TERT (mtTERT) remains poorly understood as different investigations about this mechanism contains discrepancies between their conclusions and contradictory results. It was initially observed that mtTERT increases the mitochondrial DNA (mtDNA) damage and apoptosis following treatment by H_2O_2 (Santos et al., 2003; Santos et al., 2004; Santos et al., 2006). However many other reports have also shown that mtTERT would rather display a protective role against oxidative-stress induced mtDNA damage and apoptosis. It was observed that mtTERT under oxidative stress conditions correlates with an increase in mitochondrial potential and reduction of ROS productions thus pointing to an improvement of mitochondrial function by TERT in cells subjected to oxidative stress (Ahmed et al., 2008). An increase in mitochondrial potential was previously observed in neurons. This was correlated with an increase in calcium uptake by the mitochondria as part of mechanism protecting neurons against ischaemia (Kang et al., 2004). Other recent investigations emphasize the role of mtTERT in the protection against oxidative stress. In 2009 it was reported that TERT translocation to mitochondria follows a classical pathway of proteins imported into mitochondria. Indeed it was observed that TERT translocated to mitochondrial matrix through the translocase of outer membrane (TOM) and translocase of inner membrane (TIM) complexes (Haendeler et al., 2009). Once in the matrix, it was shown that TERT can bind to mtDNA through the coding regions of the NADH:ubiquinone oxidoreductase subunits 1 and 2. This interaction between TERT and mtDNA appears to be able to protect it against ethidium-bromide induced DNA damage (Figure 2). In addition to its binding to mtDNA it was observed that cells overexpressing TERT displays an enhanced complex I activity while it reduces the ROS production induced by ethidium-bromide treatment. It is important to note here the interesting ability of TERT to bind to the loci of subunits 1 and 2 of the NADH:ubiquinone oxidoreductase which is the complex I of mitochondrial electron transport chain. One may postulate that TERT binding to these loci may enhance the gene transcription of these subunits in order to facilitate and improve mitochondrial respiration. Such a finding deserve further investigation in order to elucidate the correlation between the mtDNA-associated TERT and the increase in complex I activity. Moreover this mechanism of protection is directly correlated with the ability of TERT to localize in the mitochondria given that a construct of TERT targeted specifically to mitochondria enhanced the protective effect seen previously with TERT wild-type. The authors also have shown that the reverse transcriptase of TERT seems to be required in order to fulfill this protective role against oxidative stress (Haendeler et al., 2009). However the requirement of the reverse transcriptase activity of TERT in this protective role still remains highly controversial as no detailed mechanism has been clearly demonstrated. Nonetheless, telomerase activity has been detected in mitochondrial extracts and the binding of TERT to mtDNA suggests that the reverse transcriptase activity may play an important role in protecting mtDNA. As a consequence, it is possible to extrapolate that mtTERT can display more than one function in mitochondria and that some of them require a catalytically active telomerase (binding to mtDNA) while others may only require TERT subunit (improvement of mitochondrial function, protection against cell death) (Saretzki, 2009).

Another recent report also confirmed the role of TERT as a modulator of ROS production (Indran et al., 2010, 2011). Indeed it was observed that TERT overexpression induces reduction of basal levels of ROS and inhibits the ROS production induced by oxidative stress (Figure 2). This investigation also showed that the antioxidant function of TERT may be linked to an increase in the ratio of reduced glutathione to oxidized glutathione in addition to an improved recovery of the peroxiredoxin in its reduced state. As a substantiation of the previous results we mentioned earlier, the authors of this study were able to show that TERT induces an increase in complex IV activity (cytochrome c oxidase) (Indran et al., 2011). In the meantime it was also confirmed that these cells overexpressing TERT display a higher resistance to H_2O_2-induced apoptosis.

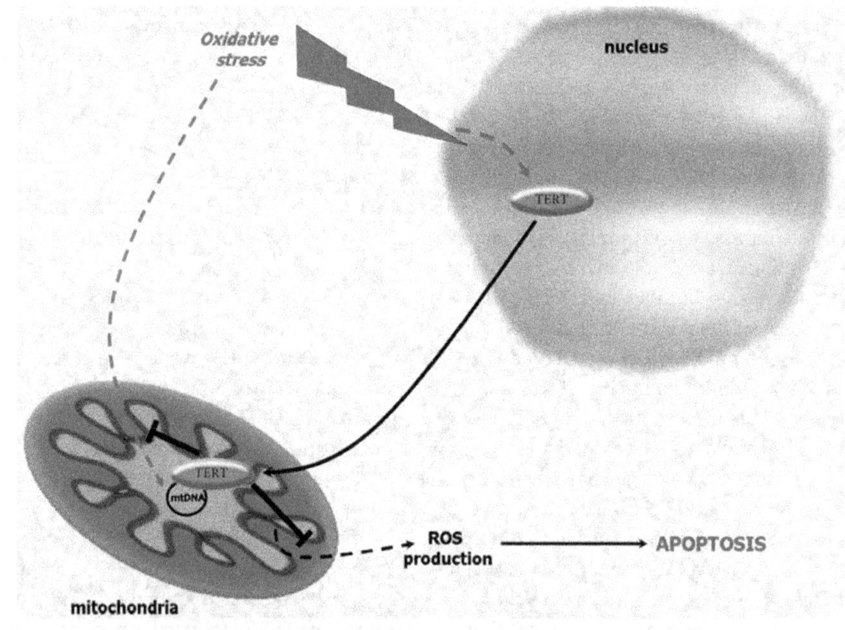

Environmental stress such as oxidative stress has been described to induce the translocation of TERT from the nucleus to the mitochondria. Once in mitochondria, TERT has been shown to interact with mtDNA and protects it against oxidative-stress-induced DNA damage. Mitochondrial TERT is also able to modulate ROS production thus promoting cell survival by inhibiting ROS-induced apoptosis.

Figure 2. Translocation of TERT into mitochondria and its potential involvement in the protection against oxidative stress.

Taken together all these studies highlight an important crosstalk between the mitochondrial localization of TERT and the modulation of ROS production. While some results are contradictory, most data suggest an involvement of TERT as part of a mechanism alleviating ROS production by mitochondria thus protecting cells against oxidative stress induced damages and cell death. The discrepancies between these different studies may be explained by the different models used in the investigations and the varied experimental settings. In addition the level of oxidative stress may also be responsible for these differences as it may represent the different responses of the cells toward mild or acute oxidative stress. As a result, TERT seems to be part of a mechanism modulating ROS production and cell response to oxidative stress. Considering the important role of ROS in cell death, cell survival and in ageing, these investigations have outlined a major new function for TERT as an upstream actor regulating ROS production and mitochondrial function which may be of critical importance to determine the fate of a cell.

3.2. Relationships between TERT and apoptotic pathways

Many studies pointed to the anti-apoptotic role of TERT independent of its enzymatic activity. Early studies in postmitotic neurons highlighted the ability of TERT to inhibit apoptosis induced by stimuli such as amyloid-beta peptide, NMDA (*N*-methyl-D-aspartate) receptor-mediated excitotoxicity or through removal of brain-derived neurotrophic factor (BDNF) (Fu et al., 2002; Kang et al., 2004; Zhu et al., 2000). Additional studies demonstrated the ability of TERT to antagonize apoptosis induced by topoisomerase inhibitors in PC12 cell line or by oxidative stress in lymphocytes CD4+ model (Lu et al., 2001; Luiten et al., 2003). Such results illustrating the protective role of TERT against ROS were confirmed later in other models (Ahmed et al., 2008; Haendeler et al., 2009; Indran et al., 2011). Although most of these early investigations did not explore the involvement of the reverse transcriptase activity of TERT in this anti-apoptotic mechanism, these results had already outlined a potential function of TERT unrelated to its enzymatic activity and ability to lengthen telomeres (Sung et al., 2005). This was further elucidated following the discovery of TERT's pro-tumorigenic function which is independent of its ability to maintain telomeres (Stewart et al., 2002).

The anti-apoptotic effect of TERT has been related to an inhibition of the mitochondrial pathway of apoptosis as it was described to inhibit the major hallmarks of the intrinsic pathway i.e., the translocation of Bax to mitochondria, the decrease in mitochondrial potential and the release of cytochrome c (Indran et al., 2011) (Figure 3). This effect was observed using a dominant negative form of TERT which resulted in an enhancement of apoptosis induced by sodium butyrate (Xi et al., 2006). Other studies highlighted the role of TERT as an antagonist of the intrinsic pathway of apoptosis. Indeed it was observed that TERT overexpression inhibits Bcl-2 dependent apoptosis (Del Bufalo et al., 2005). In this study, TERT function directly supported the anti-apoptotic role of Bcl-2, which showed that the requirement of its reverse transcriptase activity is unnecessary (Figure 3). It would be of interest to study this potential aspect on improving the survival function of Bcl-2 involved in the anti-apoptotic role of TERT, as it was described earlier that Bcl-2 itself appears to be able to regulate telomerase activity (Mandal and Kumar, 1997). As we discussed previously about the function of TERT in modulating ROS production, it is also important to note that Bcl-2 has been described as an important modulator of ROS production by mitochondria (Chen and Pervaiz, 2007, 2010; Low et al., 2011; Velaithan et al., 2011). Using these findings we can extrapolate that TERT may interact directly or indirectly with Bcl-2 and promote its anti-apoptotic function and modulate or block its pro-oxidant role as well.

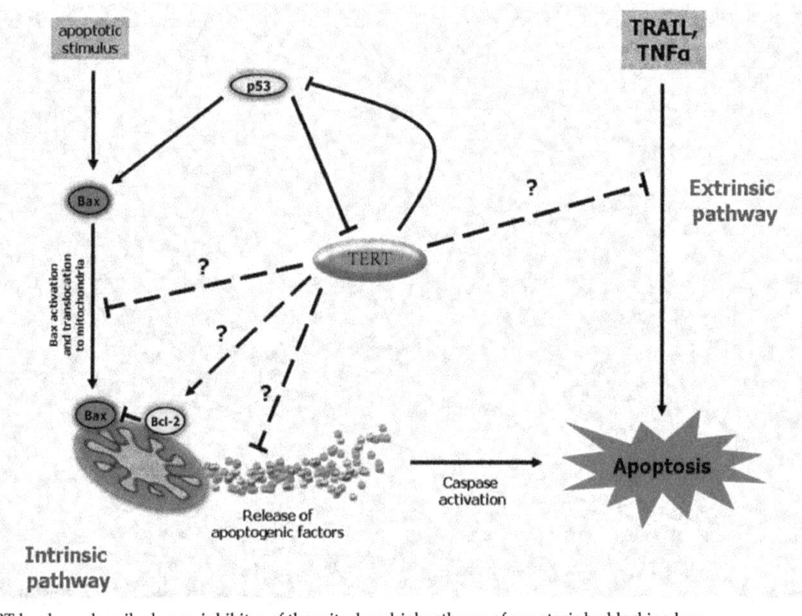

TERT has been described as an inhibitor of the mitochondrial pathway of apoptosis by blocking key events of this pathway such as Bax translocation to mitochondria and release of apoptogenic factors such as cytochrome c; however the mechanism by which TERT inhibits these events remains poorly understood. TERT has been also described as an inhibitor of the extrinsic pathway of apoptosis by blocking cell death induced by TRAIL and TNFα.

Figure 3. Relationship between TERT and the intrinsic and extrinsic pathways of apoptosis.

Other investigations have also demonstrated the ability of TERT to block the intrinsic pathway of apoptosis . The knock-down of TERT has been shown to increase the sensitivity of cancer cell lines (HeLa and HCT116) to treatments such as cisplatin, etoposide, mitomycin C and ROS mainly by facilitating the conformational activation of Bax which is the major effector of the mitochondrial pathway of apoptosis (Massard et al., 2006). This sensitization observed following TERT silencing was rescued by overexpression of Bcl-2 which constituted a hallmark of TERT contribution to the mitochondrial pathway. More recently it was also depicted in a human pancreatic cancer cell model that the silencing of TERT led to growth inhibition which associated with a decrease of Bcl-2 and cyclooxygenase 2 levels thus further deepening the connection between mitochondria, Bcl-2 and TERT (Zhong et al., 2010).

In addition other studies also confirmed the anti-apoptotic role of TERT in apoptosis induced by other stimuli such as 15-deoxy-$\Delta^{12,14}$-prostaglandin J2 (15d-PGJ2) which kills cells through induction of ROS production (Kanunfre et al., 2004; Shin et al., 2009). Interestingly it was observed that 15d-PGJ2 treatment induces TERT downregulation which seems to be an important feature of 15d-PGJ2-mediated cell death and may outline the anti-apoptotic function of TERT (Moriai et al., 2009).

Taken together, these results highlight an important role of TERT as an antagonist of the intrinsic pathway of apoptosis. This protective role does not seem to be linked to TERT ability to elongate telomeres. Although the death mechanisms induced by telomere attrition are mostly linked to the DNA repair machinery, we have described above many studies showing TERT inhibits apoptosis induced by a wide range of stimuli which are not necessarily related to the induction of DNA damage signalling. Moreover most of the effects on apoptosis sensitization occur in a short time following silencing of TERT expression (Massard et al., 2006) which does not match the timing required for a mechanism involving telomere shortening.

While most of the studies related to the involvement of TERT in apoptosis regulation pointed toward a main role of TERT as a modulator of the intrinsic pathway, several investigations also highlighted a potential role in the extrinsic pathway or receptor pathway of apoptosis. Indeed it was observed that TERT inhibits cell death induced by TNF-α and TRAIL but does not protect against etoposide and cisplatin (Dudognon et al., 2004) (Figure 3). Of note, this work also showed that the blockade of the extrinsic pathway was independent of TERT ability to maintain telomere length. It was later confirmed in another publication describing that knock down of TERT sensitizes cells to TRAIL-induced cell death (Zhang et al., 2010). More recently, it was demonstrated that TERT inhibits TNFα induced cell death by blocking the ROS-induced signalling pathways which in turn activated the downstream TNFα signalling (Mattiussi et al., 2012). Nevertheless these results remain controversial and are in contradiction to other published work. Massard and colleagues showed that TERT silencing does not affect cell sensitivity to CD95/Fas-mediated cell death (Massard et al., 2006). These discrepancies may be explained by the differences in the models used in the studies. Indeed while CD95/Fas ligand, TNF-α and TRAIL are inducers of the receptor pathway of apoptosis, the signalling pathways involved downstream are not exactly the same which may contribute to the differences between these experimental results. In addition, it may also suggest that there is a crosstalk between extrinsic and intrinsic pathways of apoptosis (Li et al., 1998) in which the mitochondrial pathway can act as an amplification loop to execute the response to stimulate the receptor pathway of apoptosis. In some cells this amplification system is essential for the total completion of the response to the receptor pathway of initiating apoptosis. As a consequence, this crosstalk between extrinsic and intrinsic pathways may help explain that in some models, TERT inhibits extrinsic pathways of apoptosis whereas in other models, it cannot fulfil its anti-apoptotic role and thus need not require the mitochondrial pathway amplification system. Another possible explanation may involve the differences in the p53 status of the cells used in these studies which often lead to different responses toward apoptotic stimuli. Taken together these results emphasize the role displayed by TERT in apoptosis regulation and more specifically in the modulation of the mitochondrial pathway of apoptosis which is in line with its ability to translocate and localize to this organelle and its capacity to modulate and protect mitochondrial functions.

3.3. Relationship between TERT and p53-dependent apoptosis

Considering the major role of the tumor suppressor p53 in the response to DNA damage and the ability of TERT to induce cell cycle arrest, apoptosis and senescence when telomeres reach a critical short length, it is important to question the relationship between TERT and the regulation of p53-mediated apoptosis (Beliveau and Yaswen, 2007; Martinez and Blasco, 2011; Vogelstein et al., 2000). Previously, it was shown that p53 is able to downregulate TERT (Kanaya et al., 2000; Xu et al., 2000). While the potential connection between the anti-apoptotic role of TERT and p53-dependent apoptosis still remains poorly understood, some results published within the past ten years may constitute as a starting point to explore this question.

Other studies investigating the ability of TERT to inhibit the mitochondrial pathway of apoptosis as well as the role of p53, concluded that p53 was not involved in this mechanism (Del Bufalo et al., 2005; Massard et al., 2006). Nevertheless, it has also been demonstrated that TERT overexpression blocked the p53-dependent apoptosis induced by 5-flurouracile, mitomycin C or activation of a temperature sensitive p53 (Rahman et al., 2005). Besides, the authors were able to show that a catalytically inactive TERT displayed an anti-apoptotic effect thus confirming a real extra-telomeric function of TERT as an antagonist of p53-mediated apoptosis. Such an inhibition of the p53-dependent apoptosis was described recently by the ability of TERT to induce basic fibroblast growth factor (bFGF) which in turn lead to a decrease in activation of p53 under DNA damage conditions (Jin et al., 2010). The induction of bFGF by TERT was independent of its reverse transcriptase activity as the catalytically inactive TERT mutant was also able to display the same response and block the DNA damage response. The results of this last study may be complementary to an earlier study highlighting a mutual regulation between p53 and TERT. Indeed it has been published previously that TERT knock down induces an increase in p53 and p21 levels (Lai et al., 2007). These results seem to outline a potential role of TERT in the regulation of its own factors which may then constitute a feedback loop in which TERT level may determine the regulation (Figure 3). As a consequence of this feedback loop, TERT appears to be able to control the level of p53 and antagonizes the p53-dependent apoptosis. Furthermore it has been observed that oxidative stress induced by hypoxia (HIF1-α upregulation) in myocardial tissues of young rats lead to an increase in p53 level. This is associated with a dramatic decrease of TERT level which then correlates with an increase in apoptotic cells in the tissue (Cataldi et al., 2009). It is also important to note that this mechanism was mostly described in myocardial tissues of young rats whereas it was less pronounced in the tissues of older rats likely due to the lower level of TERT expression. Taken together these results emphasize the important role of the HIF1-α/p53 axis in ageing as a consequence from the oxidative stress, cell death and repression of TERT expression. This mechanism while initially a tumor suppressing system may then in turn become highly tumorigenic in case of p53 mutation leading to an increase in genomic instability. Another surprising report showed that p53 and TERT were important in the mechanism known as herpes simplex virus dependent apoptosis (HDAP) specifically in the response to the viral oncoprotein E6 from human papillomavirus HPV16 and HPV18 (Nguyen et al., 2007). However the study reported that the HDAP mediated by E6 is linked to a repression of p53 while concomitantly increasing the level of TERT. This study is one of the few to report a pro-apoptotic role of the catalytic subunit TERT compared to many others highlighting its anti-apoptotic function. In addition the repression of p53 is commonly known to be associated with a higher resistance to apoptosis induced by DNA damage (Vogelstein et al., 2000). Nevertheless this mechanism of response to viral infection by HPV may outline a wider function of TERT in apoptosis regulation, which was previously reported as a "switch-like" role between life and death depending on the stress inflicted to the cells. These results emphasize a plausible link between p53-dependent apoptosis and TERT that warrants further investigation. The ability of p53 to induce apoptosis through induction of pro-apoptotic proteins such as Bax, Noxa, Fas and increase of ROS production has been well described (Vogelstein et al., 2000). On the other hand, the TERT ability to modulate p53 level as part of a mechanism of mutual regulation has been also documented previously (Jin et al., 2010; Lai et al., 2007). As a consequence, these results point toward a potential relationship between these two proteins which indicates the capacity of TERT to modulate p53-dependent apoptosis in response to a wide range of stimuli thus reflecting the ability of TERT to antagonize the p53-dependent apoptosis (Rahman et al., 2005).

4. Concluding remarks

The investigations about the extra-telomeric functions of the catalytic subunit of telomerase, TERT in the modulation of cell death has been documented in the past 10 years and has offered new insights concerning the role of telomerase in cell biology and signaling. These new findings on the supplementary role of TERT suggest that the catalytic subunit of telomerase may modulate the mitochondrial function and apoptotic cell death. This implies that TERT displays role(s) beyond the ability to lengthen telomeres and it is of importance to improve our current knowledge about these potential extra-telomeric functions of TERT. The modulation of ROS production, mitochondrial respiration and apoptosis are indeed crucial mechanisms involved in many different diseases and play a key role in tumor progression (Antonsson, 2004; Fogg et al., 2011; Hanahan and Weinberg, 2011; Low et al., 2011; Sung et al., 2005; Vogelstein et al., 2000). However most of these extra-telomeric roles of TERT remain controversial thus highlighting the need to further study this field. While the results appear to be contradictory when some emphasize the ability of TERT to prevent apoptosis while others showed the ability of TERT to enhance apoptosis (Saretzki, 2009), we must take into account the differences between the models used in these studies as well as the experimental settings. Furthermore among the studies showing the anti-apoptotic effect of TERT, the localization of TERT was not verified while it seems likely possible that nuclear TERT and mitochondrial TERT may play different roles. Indeed the mitochondrial localization of TERT has been clearly demonstrated and it was observed by Santos and colleagues that mtTERT is responsible for the sensitization to apoptosis while nuclear TERT was associated with an increase in cell survival (Santos et al., 2006). While it still needs further detailed investigations, this result could indicate that the main switch between enhancement and inhibition of cell death might be the ratio between mitochondrial and nuclear TERT. In addition, a recent work of Santos *et al.* reported that TERT can bind mitochondrial RNAs which in turn may reconstitute a reverse transcriptase activity specific to this organelle and are required for a proper mitochondrial function (Sharma et al., 2012).

As it appears that TERT clearly plays a fundamental role in mitochondria, ROS production and mitochondrial metabolism, further details concerning the mechanisms are still required to understand fully this phenomenon. Moreover, in order to determine the full extent of TERT's extra-telomeric function involved in apoptosis regulation, anti-apoptotic functions of TERT need to be methodically investigated.

Author details

Gregory Lucien Bellot
Department of Physiology, Yong Loo Lin School of Medicine,
National University of Singapore, Singapore

Xueying Wang
Department of Biochemistry, Yong Loo Lin School of Medicine,
National University of Singapore, Singapore

Acknowledgements

This work is supported by funding from the Academic Research Fund (AcRF) Tier 1 Faculty Research Committee (FRC) grants, R-183-000-295-112 and R-183-000-320-112, National University of Singapore (NUS), National University Health System (NUHS), Singapore.

5. References

Ahmed, S., Passos, J.F., Birket, M.J., Beckmann, T., Brings, S., Peters, H., Birch-Machin, M.A., von Zglinicki, T., Saretzki, G., 2008. Telomerase does not counteract telomere shortening but protects mitochondrial function under oxidative stress. Journal of cell science 121, 1046-1053.

Alder, J.K., Chen, J.J., Lancaster, L., Danoff, S., Su, S.C., Cogan, J.D., Vulto, I., Xie, M., Qi, X., Tuder, R.M., Phillips, J.A., 3rd, Lansdorp, P.M., Loyd, J.E., Armanios, M.Y., 2008. Short telomeres are a risk factor for idiopathic pulmonary fibrosis. Proceedings of the National Academy of Sciences of the United States of America 105, 13051-13056.

Antonsson, B., 2004. Mitochondria and the Bcl-2 family proteins in apoptosis signaling pathways. Molecular and cellular biochemistry 256-257, 141-155.

Beliveau, A., Yaswen, P., 2007. Soothing the watchman: telomerase reduces the p53-dependent cellular stress response. Cell Cycle 6, 1284-1287.

Benetos, A., Gardner, J.P., Zureik, M., Labat, C., Xiaobin, L., Adamopoulos, C., Temmar, M., Bean, K.E., Thomas, F., Aviv, A., 2004. Short telomeres are associated with increased carotid atherosclerosis in hypertensive subjects. Hypertension 43, 182-185.

Blackburn, E.H., Greider, C.W., Henderson, E., Lee, M.S., Shampay, J., Shippen-Lentz, D., 1989. Recognition and elongation of telomeres by telomerase. Genome / National Research Council Canada = Genome / Conseil national de recherches Canada 31, 553-560.

Blasco, M.A., 2003. Mammalian telomeres and telomerase: why they matter for cancer and aging. European journal of cell biology 82, 441-446.

Calado, R.T., Pintao, M.C., Silva, W.A., Jr., Falcao, R.P., Zago, M.A., 2002. Aplastic anaemia and telomerase RNA mutations. Lancet 360, 1608.

Calado, R.T., Regal, J.A., Kajigaya, S., Young, N.S., 2009. Erosion of telomeric single-stranded overhang in patients with aplastic anaemia carrying telomerase complex mutations. European journal of clinical investigation 39, 1025-1032.

Cataldi, A., Zara, S., Rapino, M., Zingariello, M., di Giacomo, V., Antonucci, A., 2009. p53 and telomerase control rat myocardial tissue response to hypoxia and ageing. European journal of histochemistry : EJH 53, e25.

Chen, Z.X., Pervaiz, S., 2007. Bcl-2 induces pro-oxidant state by engaging mitochondrial respiration in tumor cells. Cell death and differentiation 14, 1617-1627.

Chen, Z.X., Pervaiz, S., 2010. Involvement of cytochrome c oxidase subunits Va and Vb in the regulation of cancer cell metabolism by Bcl-2. Cell death and differentiation 17, 408-420.

Counter, C.M., 1996. The roles of telomeres and telomerase in cell life span. Mutation research 366, 45-63.

Damjanovic, A.K., Yang, Y., Glaser, R., Kiecolt-Glaser, J.K., Nguyen, H., Laskowski, B., Zou, Y., Beversdorf, D.Q., Weng, N.P., 2007. Accelerated telomere erosion is associated with a declining immune function of caregivers of Alzheimer's disease patients. J Immunol 179, 4249-4254.

de Lange, T., 2005. Shelterin: the protein complex that shapes and safeguards human telomeres. Genes & development 19, 2100-2110.

de Lange, T., 2010. How shelterin solves the telomere end-protection problem. Cold Spring Harbor symposia on quantitative biology 75, 167-177.

De Semir, D., Nosrati, M., Li, S., Kashani-Sabet, M., 2007. Telomerase: going beyond the ends. Cell Cycle 6, 546-549.

Del Bufalo, D., Rizzo, A., Trisciuoglio, D., Cardinali, G., Torrisi, M.R., Zangemeister-Wittke, U., Zupi, G., Biroccio, A., 2005. Involvement of hTERT in apoptosis induced by interference with Bcl-2 expression and function. Cell death and differentiation 12, 1429-1438.

Deng, Y., Chang, S., 2007. Role of telomeres and telomerase in genomic instability, senescence and cancer. Laboratory investigation; a journal of technical methods and pathology 87, 1071-1076.

Diotti, R., Loayza, D., 2011. Shelterin complex and associated factors at human telomeres. Nucleus 2, 119-135.

Djojosubroto, M.W., Choi, Y.S., Lee, H.W., Rudolph, K.L., 2003. Telomeres and telomerase in aging, regeneration and cancer. Molecules and cells 15, 164-175.

Dudognon, C., Pendino, F., Hillion, J., Saumet, A., Lanotte, M., Segal-Bendirdjian, E., 2004. Death receptor signaling regulatory function for telomerase: hTERT abolishes TRAIL-induced apoptosis, independently of telomere maintenance. Oncogene 23, 7469-7474.

Epel, E.S., Lin, J., Wilhelm, F.H., Wolkowitz, O.M., Cawthon, R., Adler, N.E., Dolbier, C., Mendes, W.B., Blackburn, E.H., 2006. Cell aging in relation to stress arousal and cardiovascular disease risk factors. Psychoneuroendocrinology 31, 277-287.

Fogg, V.C., Lanning, N.J., Mackeigan, J.P., 2011. Mitochondria in cancer: at the crossroads of life and death. Chinese journal of cancer 30, 526-539.

Fu, W., Lu, C., Mattson, M.P., 2002. Telomerase mediates the cell survival-promoting actions of brain-derived neurotrophic factor and secreted amyloid precursor protein in developing hippocampal neurons. The Journal of neuroscience : the official journal of the Society for Neuroscience 22, 10710-10719.

Gardner, J.P., Li, S., Srinivasan, S.R., Chen, W., Kimura, M., Lu, X., Berenson, G.S., Aviv, A., 2005. Rise in insulin resistance is associated with escalated telomere attrition. Circulation 111, 2171-2177.

Gordon, D.M., Santos, J.H., 2010. The emerging role of telomerase reverse transcriptase in mitochondrial DNA metabolism. Journal of nucleic acids 2010.

Goronzy, J.J., Fujii, H., Weyand, C.M., 2006. Telomeres, immune aging and autoimmunity. Experimental gerontology 41, 246-251.

Greider, C.W., Blackburn, E.H., 1985. Identification of a specific telomere terminal transferase activity in Tetrahymena extracts. Cell 43, 405-413.

Greider, C.W., Blackburn, E.H., 1987. The telomere terminal transferase of Tetrahymena is a ribonucleoprotein enzyme with two kinds of primer specificity. Cell 51, 887-898.

Haendeler, J., Drose, S., Buchner, N., Jakob, S., Altschmied, J., Goy, C., Spyridopoulos, I., Zeiher, A.M., Brandt, U., Dimmeler, S., 2009. Mitochondrial telomerase reverse transcriptase binds to and protects mitochondrial DNA and function from damage. Arteriosclerosis, thrombosis, and vascular biology 29, 929-935.

Hanahan, D., Weinberg, R.A., 2011. Hallmarks of cancer: the next generation. Cell 144, 646-674.

Harley, C.B., 2008. Telomerase and cancer therapeutics. Nat Rev Cancer 8, 167-179.

Hayflick, L., 1965. The Limited in Vitro Lifetime of Human Diploid Cell Strains. Experimental cell research 37, 614-636.

Hayflick, L., Moorhead, P.S., 1961. The serial cultivation of human diploid cell strains. Experimental cell research 25, 585-621.

Hiyama, E., Hiyama, K., 2007. Telomere and telomerase in stem cells. British journal of cancer 96, 1020-1024.

Indran, I.R., Hande, M.P., Pervaiz, S., 2010. Tumor cell redox state and mitochondria at the center of the non-canonical activity of telomerase reverse transcriptase. Molecular aspects of medicine 31, 21-28.

Indran, I.R., Hande, M.P., Pervaiz, S., 2011. hTERT overexpression alleviates intracellular ROS production, improves mitochondrial function, and inhibits ROS-mediated apoptosis in cancer cells. Cancer research 71, 266-276.

Jin, X., Beck, S., Sohn, Y.W., Kim, J.K., Kim, S.H., Yin, J., Pian, X., Kim, S.C., Choi, Y.J., Kim, H., 2010. Human telomerase catalytic subunit (hTERT) suppresses p53-mediated anti-apoptotic response via induction of basic fibroblast growth factor. Experimental & molecular medicine 42, 574-582.

Kanaya, T., Kyo, S., Hamada, K., Takakura, M., Kitagawa, Y., Harada, H., Inoue, M., 2000. Adenoviral expression of p53 represses telomerase activity through down-regulation of human telomerase reverse transcriptase transcription. Clinical cancer research : an official journal of the American Association for Cancer Research 6, 1239-1247.

Kang, H.J., Choi, Y.S., Hong, S.B., Kim, K.W., Woo, R.S., Won, S.J., Kim, E.J., Jeon, H.K., Jo, S.Y., Kim, T.K., Bachoo, R., Reynolds, I.J., Gwag, B.J., Lee, H.W., 2004. Ectopic expression of the catalytic subunit of telomerase protects against brain injury resulting from ischemia and NMDA-induced neurotoxicity. The Journal of neuroscience : the official journal of the Society for Neuroscience 24, 1280-1287.

Kanunfre, C.C., da Silva Freitas, J.J., Pompeia, C., Goncalves de Almeida, D.C., Cury-Boaventura, M.F., Verlengia, R., Curi, R., 2004. Ciglitizone and 15d PGJ2 induce apoptosis in Jurkat and Raji cells. International immunopharmacology 4, 1171-1185.

Keith, W.N., Thomson, C.M., Howcroft, J., Maitland, N.J., Shay, J.W., 2007. Seeding drug discovery: integrating telomerase cancer biology and cellular senescence to uncover new therapeutic opportunities in targeting cancer stem cells. Drug discovery today 12, 611-621.

Kuhlow, D., Florian, S., von Figura, G., Weimer, S., Schulz, N., Petzke, K.J., Zarse, K., Pfeiffer, A.F., Rudolph, K.L., Ristow, M., 2010. Telomerase deficiency impairs glucose metabolism and insulin secretion. Aging 2, 650-658.

Lai, S.R., Cunningham, A.P., Huynh, V.Q., Andrews, L.G., Tollefsbol, T.O., 2007. Evidence of extra-telomeric effects of hTERT and its regulation involving a feedback loop. Experimental cell research 313, 322-330.

Li, H., Zhu, H., Xu, C.J., Yuan, J., 1998. Cleavage of BID by caspase 8 mediates the mitochondrial damage in the Fas pathway of apoptosis. Cell 94, 491-501.

Liew, C.W., Holman, A., Kulkarni, R.N., 2009. The roles of telomeres and telomerase in beta-cell regeneration. Diabetes, obesity & metabolism 11 Suppl 4, 21-29.

Longhese, M.P., Anbalagan, S., Martina, M., Bonetti, D., 2012. The role of shelterin in maintaining telomere integrity. Frontiers in bioscience : a journal and virtual library 17, 1715-1728.

Low, I.C., Kang, J., Pervaiz, S., 2011. Bcl-2: a prime regulator of mitochondrial redox metabolism in cancer cells. Antioxidants & redox signaling 15, 2975-2987.

Lu, C., Fu, W., Mattson, M.P., 2001. Telomerase protects developing neurons against DNA damage-induced cell death. Brain research. Developmental brain research 131, 167-171.

Luiten, R.M., Pene, J., Yssel, H., Spits, H., 2003. Ectopic hTERT expression extends the life span of human CD4+ helper and regulatory T-cell clones and confers resistance to oxidative stress-induced apoptosis. Blood 101, 4512-4519.

Majerska, J., Sykorova, E., Fajkus, J., 2011. Non-telomeric activities of telomerase. Molecular bioSystems 7, 1013-1023.

Mandal, M., Kumar, R., 1997. Bcl-2 modulates telomerase activity. The Journal of biological chemistry 272, 14183-14187.

Martinez, P., Blasco, M.A., 2010. Role of shelterin in cancer and aging. Aging cell 9, 653-666.

Martinez, P., Blasco, M.A., 2011. Telomeric and extra-telomeric roles for telomerase and the telomere-binding proteins. Nat Rev Cancer 11, 161-176.

Mason, P.J., Wilson, D.B., Bessler, M., 2005. Dyskeratosis congenita -- a disease of dysfunctional telomere maintenance. Current molecular medicine 5, 159-170.

Massard, C., Zermati, Y., Pauleau, A.L., Larochette, N., Metivier, D., Sabatier, L., Kroemer, G., Soria, J.C., 2006. hTERT: a novel endogenous inhibitor of the mitochondrial cell death pathway. Oncogene 25, 4505-4514.

Mattiussi, M., Tilman, G., Lenglez, S., Decottignies, A., 2012. Human telomerase represses ROS-dependent cellular responses to Tumor Necrosis Factor-alpha without affecting NF-kappaB activation. Cellular signalling 24, 708-717.

Mitchell, J.R., Wood, E., Collins, K., 1999. A telomerase component is defective in the human disease dyskeratosis congenita. Nature 402, 551-555.

Moriai, M., Tsuji, N., Kobayashi, D., Kuribayashi, K., Watanabe, N., 2009. Down-regulation of hTERT expression plays an important role in 15-deoxy-Delta12,14-prostaglandin J2-induced apoptosis in cancer cells. International journal of oncology 34, 1363-1372.

Moyzis, R.K., Buckingham, J.M., Cram, L.S., Dani, M., Deaven, L.L., Jones, M.D., Meyne, J., Ratliff, R.L., Wu, J.R., 1988. A highly conserved repetitive DNA sequence, (TTAGGG)n, present at the telomeres of human chromosomes. Proceedings of the National Academy of Sciences of the United States of America 85, 6622-6626.

Nguyen, M.L., Kraft, R.M., Aubert, M., Goodwin, E., DiMaio, D., Blaho, J.A., 2007. p53 and hTERT determine sensitivity to viral apoptosis. Journal of virology 81, 12985-12995.

O'Sullivan, R.J., Karlseder, J., 2010. Telomeres: protecting chromosomes against genome instability. Nature reviews. Molecular cell biology 11, 171-181.

Oishi, T., Kigawa, J., Minagawa, Y., Shimada, M., Takahashi, M., Terakawa, N., 1998. Alteration of telomerase activity associated with development and extension of epithelial ovarian cancer. Obstetrics and gynecology 91, 568-571.

Ouellette, M.M., Wright, W.E., Shay, J.W., 2011. Targeting telomerase-expressing cancer cells. Journal of cellular and molecular medicine 15, 1433-1442.

Pirker, C., Holzmann, K., Spiegl-Kreinecker, S., Elbling, L., Thallinger, C., Pehamberger, H., Micksche, M., Berger, W., 2003. Chromosomal imbalances in primary and metastatic melanomas: over-representation of essential telomerase genes. Melanoma research 13, 483-492.

Rahman, R., Latonen, L., Wiman, K.G., 2005. hTERT antagonizes p53-induced apoptosis independently of telomerase activity. Oncogene 24, 1320-1327.

Saelens, X., Festjens, N., Vande Walle, L., van Gurp, M., van Loo, G., Vandenabeele, P., 2004. Toxic proteins released from mitochondria in cell death. Oncogene 23, 2861-2874.

Santos, J.H., Hunakova, L., Chen, Y., Bortner, C., Van Houten, B., 2003. Cell sorting experiments link persistent mitochondrial DNA damage with loss of mitochondrial membrane potential and apoptotic cell death. The Journal of biological chemistry 278, 1728-1734.

Santos, J.H., Meyer, J.N., Skorvaga, M., Annab, L.A., Van Houten, B., 2004. Mitochondrial hTERT exacerbates free-radical-mediated mtDNA damage. Aging cell 3, 399-411.

Santos, J.H., Meyer, J.N., Van Houten, B., 2006. Mitochondrial localization of telomerase as a determinant for hydrogen peroxide-induced mitochondrial DNA damage and apoptosis. Human molecular genetics 15, 1757-1768.

Saretzki, G., 2009. Telomerase, mitochondria and oxidative stress. Experimental gerontology 44, 485-492.

Saretzki, G., Murphy, M.P., von Zglinicki, T., 2003. MitoQ counteracts telomere shortening and elongates lifespan of fibroblasts under mild oxidative stress. Aging cell 2, 141-143.

Sharma, N.K., Reyes, A., Green, P., Caron, M.J., Bonini, M.G., Gordon, D.M., Holt, I.J., Santos, J.H., 2012. Human telomerase acts as a hTR-independent reverse transcriptase in mitochondria. Nucleic acids research 40, 712-725.

Shay, J.W., Bacchetti, S., 1997. A survey of telomerase activity in human cancer. Eur J Cancer 33, 787-791.

Shay, J.W., Keith, W.N., 2008. Targeting telomerase for cancer therapeutics. British journal of cancer 98, 677-683.

Shay, J.W., Wright, W.E., 2004. Telomeres are double-strand DNA breaks hidden from DNA damage responses. Molecular cell 14, 420-421.

Shay, J.W., Wright, W.E., 2011. Role of telomeres and telomerase in cancer. Seminars in cancer biology 21, 349-353.

Shin, J.S., Hong, A., Solomon, M.J., Lee, C.S., 2006. The role of telomeres and telomerase in the pathology of human cancer and aging. Pathology 38, 103-113.

Shin, S.W., Seo, C.Y., Han, H., Han, J.Y., Jeong, J.S., Kwak, J.Y., Park, J.I., 2009. 15d-PGJ2 induces apoptosis by reactive oxygen species-mediated inactivation of Akt in leukemia and colorectal cancer cells and shows in vivo antitumor activity. Clinical cancer research : an official journal of the American Association for Cancer Research 15, 5414-5425.

Shtessel, L., Ahmed, S., 2011. Telomere dysfunction in human bone marrow failure syndromes. Nucleus 2, 24-29.

Stewart, S.A., Hahn, W.C., O'Connor, B.F., Banner, E.N., Lundberg, A.S., Modha, P., Mizuno, H., Brooks, M.W., Fleming, M., Zimonjic, D.B., Popescu, N.C., Weinberg, R.A., 2002. Telomerase contributes to tumorigenesis by a telomere length-independent mechanism. Proceedings of the National Academy of Sciences 99, 12606-12611.

Sung, Y.H., Choi, Y.S., Cheong, C., Lee, H.W., 2005. The pleiotropy of telomerase against cell death. Molecules and cells 19, 303-309.

Takai, H., Smogorzewska, A., de Lange, T., 2003. DNA damage foci at dysfunctional telomeres. Current biology : CB 13, 1549-1556.

Tsakiri, K.D., Cronkhite, J.T., Kuan, P.J., Xing, C., Raghu, G., Weissler, J.C., Rosenblatt, R.L., Shay, J.W., Garcia, C.K., 2007. Adult-onset pulmonary fibrosis caused by mutations in telomerase. Proceedings of the National Academy of Sciences of the United States of America 104, 7552-7557.

Velaithan, R., Kang, J., Hirpara, J.L., Loh, T., Goh, B.C., Le Bras, M., Brenner, C., Clement, M.V., Pervaiz, S., 2011. The small GTPase Rac1 is a novel binding partner of Bcl-2 and stabilizes its antiapoptotic activity. Blood 117, 6214-6226.

Vogelstein, B., Lane, D., Levine, A.J., 2000. Surfing the p53 network. Nature 408, 307-310.

Vulliamy, T., Marrone, A., Dokal, I., Mason, P.J., 2002. Association between aplastic anaemia and mutations in telomerase RNA. Lancet 359, 2168-2170.

Wright, W.E., Piatyszek, M.A., Rainey, W.E., Byrd, W., Shay, J.W., 1996. Telomerase activity in human germline and embryonic tissues and cells. Developmental genetics 18, 173-179.

Wyatt, H.D., West, S.C., Beattie, T.L., 2010. InTERTpreting telomerase structure and function. Nucleic acids research 38, 5609-5622.

Xi, L., Chen, G., Zhou, J., Xu, G., Wang, S., Wu, P., Zhu, T., Zhang, A., Yang, W., Xu, Q., Lu, Y., Ma, D., 2006. Inhibition of telomerase enhances apoptosis induced by sodium butyrate via mitochondrial pathway. Apoptosis : an international journal on programmed cell death 11, 789-798.

Xu, D., Wang, Q., Gruber, A., Bjorkholm, M., Chen, Z., Zaid, A., Selivanova, G., Peterson, C., Wiman, K.G., Pisa, P., 2000. Downregulation of telomerase reverse transcriptase mRNA expression by wild type p53 in human tumor cells. Oncogene 19, 5123-5133.

Yuan, S., Yu, X., Asara, J.M., Heuser, J.E., Ludtke, S.J., Akey, C.W., 2011. The holo-apoptosome: activation of procaspase-9 and interactions with caspase-3. Structure 19, 1084-1096.

Zhang, R.G., Zhao, J.J., Yang, L.Q., Yang, S.M., Wang, R.Q., Chen, W.S., Peng, G.Y., Fang, D.C., 2010. RNA interference-mediated hTERT inhibition enhances TRAIL-induced apoptosis in resistant hepatocellular carcinoma cells. Oncology reports 23, 1013-1019.

Zhong, Y.Q., Xia, Z.S., Fu, Y.R., Zhu, Z.H., 2010. Knockdown of hTERT by SiRNA suppresses growth of Capan-2 human pancreatic cancer cell via the inhibition of expressions of Bcl-2 and COX-2. Journal of digestive diseases 11, 176-184.

Zhu, H., Fu, W., Mattson, M.P., 2000. The catalytic subunit of telomerase protects neurons against amyloid beta-peptide-induced apoptosis. Journal of neurochemistry 75, 117-124.

Chapter 7

Programmed Cell Death in T Cell Development

Qian Nancy Hu and Troy A. Baldwin

Additional information is available at the end of the chapter

1. Introduction

The mammalian immune system is a complex network of many cell types and proteins that collectively coordinate a protective response against foreign entities. The immune system can be divided into two broad categories: innate and adaptive immunity. Adaptive immunity, which is primarily mediated by T and B lymphocytes, first arose in jawed vertebrates and has several distinct features from the more ancient innate immune system (Pancer and Cooper 2006). While innate immunity is characterized by non-specific recognition of conserved molecular patterns leading to a rapid effector response, the adaptive immune response is delayed by the required expansion of lymphocytes bearing receptors specific for a particular antigen. After the primary immune response, a small fraction of activated lymphocytes remain as memory cells, which respond to subsequent encounters with the same antigen in a more rapid and robust manner.

Due to the enormous diversity of antigens to which the adaptive immune system must respond, generation of antigen receptors cannot occur on a one gene-one protein basis. Instead, diversity is achieved through recombination of genetic segments, nucleotide additions and deletions, and pairing of different chains to form the complete antigen receptor. For example, of a theoretical 10^{15} T cell receptor (TCR) specificities in humans, only a small fraction is represented by circulating T cells (Arstila et al. 1999). Most lymphocytes do not complete development and are eliminated by programmed cell death. This can occur at several checkpoints both independent and dependent of the antigen receptor specificity. T and B cell development are analogous; both ultimately require the generation of an antigen receptor that can bind self-antigen with an affinity just high enough to enable survival and maturation. Lymphocytes bearing receptors with excessively high affinity for self-antigen are eliminated. This chapter focuses on the role of programmed cell death during T cell development in the thymus. Recent advances in the field reveal a complex and elegant system designed to select for immunocompetent and self-tolerant T cells.

2. TCR-independent T cell development

T cell development occurs in the thymus, a bilobed organ composed of an outer capsule, a peripheral cortex, and a central medulla. T cell progenitors from the bone marrow or fetal liver seed the thymic cortex as double negative (DN) thymocytes, so called because they lack expression of CD4 and CD8 co-receptors. Murine DN thymocytes progress through four stages of development defined by differential expression of the proteins CD44 and CD25 (Godfrey et al. 1993) (**Figure 1**). The earliest progenitors that seed the thymus are termed DN1 ($CD44^{+}CD25^{-}$). T cell lineage commitment initiates rearrangement of the TCRβ chain by recombination activating gene-1 and -2 (Rag-1 and Rag-2) enzymes at the DN2 stage ($CD44^{+}CD25^{-}$), which continues into the DN3 stage ($CD44^{-}CD25^{+}$) (Livak et al. 1999). Rearrangement of the TCRγ and δ loci are also evident in DN3 thymocytes and it is at this stage that the γδ T cell lineage diverges from conventional αβ T cells (MacDonald et al. 2001). In this chapter, we focus on the role of apoptosis in αβ T cell development.

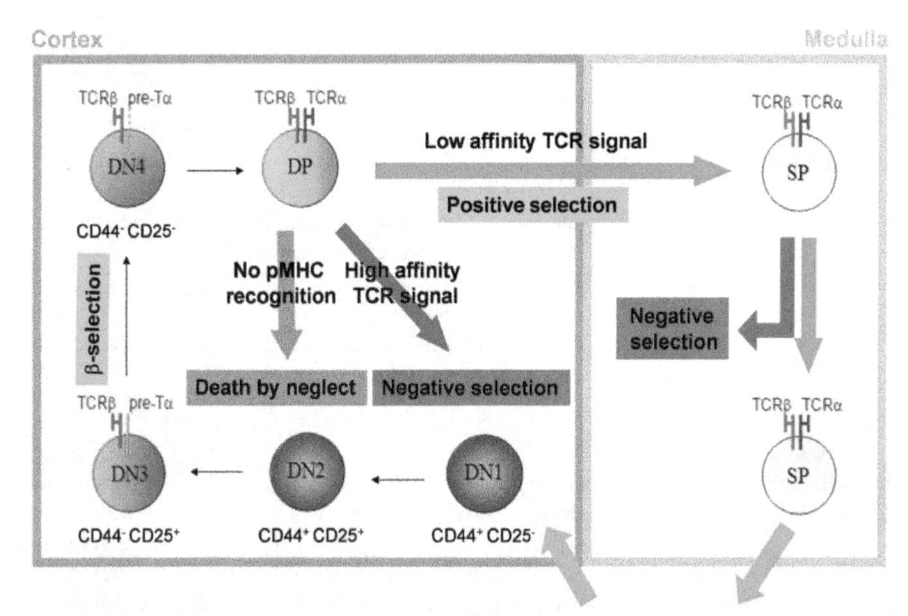

Figure 1. Conventional T cell development in the thymus.

T cell progenitors enter the thymus on the cortical side of the cortico-medullary junction as DN thymocytes, lacking expression of CD4 and CD8 co-receptors. DN thymocytes progress through four stages of development defined by differential expression of CD44 and CD25. Only thymocytes expressing a functional TCRβ chain survive the β-selection checkpoint at the DN3 to DN4 transition and are permitted to continue development into $CD4^{+}CD8^{+}$ DP thymocytes. DP thymocytes have three fates depending on the affinity of its TCR for self-pMHC: death by neglect, negative selection, or positive selection. Positively selected thymocytes differentiate into $CD4^{+}CD8^{-}$ or $CD4^{-}CD8^{+}$ SP cells and migrate to the medulla, where negative selection against tissue-restricted antigens occurs. SP thymocytes that survive this process enter the peripheral T cell repertoire.

While the mature αβTCR is composed of one α and β chain each, the TCRβ chain is initially paired with an invariant pre-TCRα chain, together forming the pre-TCR (Saint-Ruf et al. 1994). Signaling through the pre-TCR or mature TCR is mediated by an associated protein complex that contains a CD3γ/CD3ε heterodimer, CD3δ/CD3ε heterodimer, and CD3ζ homodimer (**Figure 2**). The TCR only recognizes peptide antigen when presented on major histocompatibility complex (MHC) molecules. A common method of artificially stimulating thymocytes in vitro is using agonist antibodies against CD3ε and the co-stimulatory molecule CD28. In contrast to the mature TCR, interaction between the pre-TCR and peptide-MHC (pMHC) is not required for signaling (Irving et al. 1998). Rather, translocation of pre-TCR complexes to lipid rafts in the plasma membrane has been proposed to provide a platform for association with signaling proteins (Saint-Ruf et al. 2000). Pre-TCR signaling is required for allelic exclusion at the TCRβ locus, survival, differentiation into DN4 ($CD44^{-}$ $CD25^{-}$) thymocytes, proliferation and subsequent differentiation into $CD4^{+}CD8^{+}$ (DP) thymocytes, and initiation of TCRα rearrangement (Michie and Zuniga-Pflucker 2002). This has been demonstrated by genetic ablation of Rag-2, pre-TCRα, and numerous downstream signaling components (review of pre-TCR and TCR signaling pathways in thymocytes in references (Michie and Zuniga-Pflucker 2002; Starr et al. 2003)). Since this process selects for thymocytes with functional TCRβ rearrangements, it is referred to as the β-selection checkpoint.

Rearrangement of the TCRα locus occurs at low levels in DN4 thymocytes but does not occur at full scale until the DP stage (Hernandez-Munain et al. 1999). DP thymocytes have a lifespan of 3-4 days (Egerton et al. 1990), during which time multiple recombination events can occur at each TCRα allele until the generation of an αβTCR that engages self-pMHC expressed on cortical thymic epithelial cells. The majority of DP thymocytes do not express a functional αβTCR or do not engage self-pMHC within this time window. Thymocytes that fail this self-MHC restriction checkpoint undergo "death by neglect" to maximize the utility of the T cell repertoire.

2.1. Apoptosis in pre-β-selection thymocytes

The thymic microenvironment provides signals to developing thymocytes through cell-cell interactions and cytokines. Two key cell surface receptors important for survival of pre-β-selection thymocytes are the IL-7 receptor and Notch.

IL-7 is a cytokine produced by thymic epithelial cells that signals through a receptor consisting of an IL-7Rα chain and a γ_c chain, resulting in the activation of a variety of signaling pathways. IL-7 is known to promote survival and proliferation but this chapter will focus only on the role of IL-7 signaling in thymocyte survival. Early studies showed that

IL-7 increases the viability of DN1, DN2, and DN3 thymocytes cultured *in vitro* (Godfrey et al. 1993; Kim et al. 1998). *In vivo* studies using IL-7R$\alpha^{-/-}$ mice revealed a deficit of CD25^{+} DN (i.e. DN2 and DN3) thymocytes, suggesting that IL-7 is needed for the transition to and/or at these stages (Peschon et al. 1994; Maraskovsky et al. 1997). Later studies that examined sorted DN populations confirm that the DN2 and DN3 subsets have the highest expression of IL-7Rα (Yu et al. 2004) and responsiveness to IL-7 stimulation (Van De Wiele et al. 2004). IL-7 promotes thymocyte survival through both positive regulation of anti-apoptotic members and negative regulation of pro-apoptotic members of the B cell lymphoma (Bcl) 2 family (for a summary of the role of Bcl-2 family members in T cell development, refer to **Table 1**). For example, IL-7 stimulation of CD3ε^{-} DN thymocytes induces expression of the anti-apoptotic protein Bcl-2 (von Freeden-Jeffry et al. 1997; Kim et al. 1998) and overexpression of Bcl-2 in IL-7R$\alpha^{-/-}$ (Maraskovsky et al. 1997; Khaled et al. 2002) or $\gamma_c^{-/-}$ (Kondo et al. 1997) mice is sufficient to rescue development. Additionally, myeloid cell leukemia sequence 1 (Mcl-1), another anti-apoptotic Bcl-2 family member, is induced by IL-7 and can promote DN2 survival (Opferman et al. 2003). While a role for the pro-apoptotic protein Bcl-2 homologous antagonist/killer (Bak) is controversial (Khaled et al. 2002; Dunkle et al. 2010), a role for Bcl-2-associated X protein (Bax) in death by IL-7 starvation is more established. IL-7 stimulation of DN thymocytes reduces Bax expression, while IL-7 withdrawal causes mitochondrial translocation of Bax in a DN2-derived cell line called D1 (Kim et al. 1998; Khaled et al. 1999). Furthermore, Bax deficiency partially rescues DN thymocyte death in IL-7R$\alpha^{-/-}$ mice (Khaled et al. 2002). IL-7 signaling is thought to promote DN thymocyte survival in part due to activation of the phosphoinositide 3-kinase (PI3K)/Akt pathway. Akt is activated following recruitment to the plasma membrane molecule phosphatidylinositol (3,4,5)-triphosphate (PIP_3), whose production is mediated by PI3K and inhibited by phosphatase and tensin homolog (PTEN) (Song et al. 2005). Consistent with high IL-7 responsiveness, DN2 thymocytes are especially sensitive to death when treated with PI3K inhibitor (Khaled et al. 2002). One substrate of Akt is the pro-apoptotic protein Bcl-2-associated death promoter (Bad), which is sequestered in the cytosol when phosphorylated by Akt. IL-7 stimulation of DN2 and DN3 thymocytes increases Bad phosphorylation, which in D1 cells was shown to be Akt-dependent (Khaled et al. 2002; Li et al. 2004). In contrast, IL-7 withdrawal results in translocation of Bad to mitochondria, where it inhibits Bcl-2 to promote Bax activation. While the pro-apoptotic Bcl-2 member Bcl-2-interacting mediator of cell death (Bim) is another known target of Akt, Bim expression in DN thymocytes is not regulated by IL-7 and Bim deficiency cannot rescue the DN2 and DN3 blocks in adult IL-7R$\alpha^{-/-}$ mice (Khaled et al. 2002; Pellegrini et al. 2004). Additionally, IL-7 has been shown to promote survival of immature (CD34^{+}) human thymocytes through activation of the PI3K/Akt pathway, though this marker does not distinguish between pre- and post-β-selection DN cells (Pallard et al. 1999). Taken together, these results indicate that IL-7 is critical for pre-β-selection thymocyte survival largely through regulation of Bcl-2 family members and the intrinsic apoptosis pathway. Consistent with this, DN death in the absence of IL-7 is characterized by DNA fragmentation and annexin V binding (Kim et al. 1998). However, the caspase inhibitors z-VAD and z-DEVD abrogate DNA fragmentation but do not prevent death, indicating the contribution of a non-apoptotic cell death pathway.

In addition to producing cytokines, thymic epithelial cells regulate thymocyte development through direct interactions. Interactions between Notch proteins and ligands of the Delta-like and Jagged families are critical at several stages of T cell development. In mice, ligand binding induces proteolytic release of the Notch intracellular domain, which translocates to the nucleus and activates the transcriptional regulator recombination signal binding protein for immunoglobulin kappa J region (RBP-J), leading to transcription of target genes. Inducible deletion of Notch1 or RBP-J in bone marrow precursors results in loss of thymocytes as early as the DN1 stage and development of thymic precursors into B cells (Wilson et al. 2001; Han et al. 2002). While Notch1 is critical for early T cell lineage commitment, little is known about the role of Notch1 in survival of pre-β-selection thymocytes. Delta-like4 is the non-redundant Notch1 ligand required for T cell development *in vivo* (Hozumi et al. 2008; Koch et al. 2008) but ectopic expression of either Delta-like1 or Delta-like4 on OP9 bone marrow-derived stromal cells can support T cell development *in vitro* (Schmitt and Zuniga-Pflucker 2002; Hozumi et al. 2004). Culture of pre-β-selection murine ($Rag2^{-/-}$) or human ($CD3\varepsilon^{-}$) DN thymocytes in the presence of Delta-like1 or Delta-like4 results in a decreased frequency of cell death (Ciofani et al. 2004; Magri et al. 2009). Notch1 promotes survival of $Rag2^{-/-}$ DN3 thymocytes by maintaining glucose metabolism in an Akt-dependent manner (Ciofani and Zuniga-Pflucker 2005). Thus, it is hypothesized that Notch1 signaling generates a metabolic environment permissive for β-selection. More work is needed to characterize the anti-apoptotic aspect of Notch1 signaling during early T cell development *in vivo*.

2.2. Life and death at the β-selection checkpoint

While IL-7 is critical for T cell development up to the DN3 stage, several studies have found little to no role for IL-7 signaling in thymocyte survival during β-selection. This is due to the fact that post-β-selection DN3 and DN4 thymocytes have decreased expression of IL-7Rα and subsequently reduced responsiveness to IL-7 compared to pre-β-selection stages (Van De Wiele et al. 2004; Yu et al. 2004; Van de Wiele et al. 2007; Teague et al. 2010). Interestingly, transgenic IL-7Rα expression ultimately impairs differentiation into DP thymocytes, suggesting that progression through β-selection is impeded by IL-7 signaling (Yu et al. 2004).

Thymocyte survival during β-selection critically depends on signals downstream of the pre-TCR. Consistent with this, activation of caspase-3 is apparent in $Rag1^{-/-}$ DN3 thymocytes and is abrogated upon anti-CD3ε stimulation (Mandal et al. 2005). In contrast to its role in promoting survival of pre-β-selection DN3 thymocytes, Notch1 signaling has been proposed to eliminate cells that fail to rearrange the TCRβ chain (Wolfer et al. 2002). Interestingly, Notch1 expression is induced by the transcription factor E2A, which is also implicated in elimination of $TCR\beta^{-}$ thymocytes (Michie and Zuniga-Pflucker 2002). In support of this, $E2A^{-/-}$ mice develop lymphoma and ectopic expression of E2A induces death but not cell cycle arrest in lymphoma cells (Engel and Murre 1999). Pre-TCR signaling negatively regulates E2A activity and Notch1 expression, which may be one mechanism of promoting survival of thymocytes that express functional TCRβ chains (Yashiro-Ohtani et al. 2009).

Contrary to these reports, constitutive Notch1 activity in Rag2$^{-/-}$ mice does not increase thymocyte apoptosis before or after anti-CD3ε stimulation, arguing against the idea that Notch1 promotes death of TCRβ^{-} cells during β-selection (Huang et al. 2003). Thus, the role of Notch1 in thymocyte survival at the β-selection checkpoint requires further clarification.

As in pre-β-selection thymocytes, the PI3K/Akt pathway plays an important role in thymocyte survival during β-selection. Pre-TCR signaling during β-selection induces Akt activation in a Notch1-independent manner (Ciofani and Zuniga-Pflucker 2005; Mao et al. 2007). Mice deficient in PI3K or Akt show a partial block at the DN3 to DN4 transition, while deletion of PTEN or expression of constitutively active Akt rescues β-selection in Rag2$^{-/-}$ mice (Rodriguez-Borlado et al. 2003; Hagenbeek et al. 2004; Mao et al. 2007). In PTEN-deficient Rag2$^{-/-}$ mice, unchecked Akt activation allows the survival and expansion of abnormal TCRβ^{-} cells (Hagenbeek et al. 2004). Conversely, deletion of Akt results in increased DN thymocyte death (Juntilla et al. 2007; Mao et al. 2007). Activated Akt controls the expression of several downstream apoptotic proteins. For example, through phosphorylation and nuclear exclusion of the transcription factor forkhead box O3a (FoxO3a), pre-TCR-induced Akt promotes transcriptional downregulation of Bim (Mandal et al. 2008). Survival during β-selection critically depends on pre-TCR-induced inactivation of the tumour suppressor p53, which would normally induce apoptosis in response to double-stranded DNA breaks such as those caused by TCRβ recombination (Jiang et al. 1996; Haks et al. 1999). Using pre-TCRα$^{-/-}$ DN3 thymocytes, which can undergo TCRβ recombination but not pre-TCR signaling, it was found that p53 expression remains high and directly activates transcription of the pro-apoptotic protein Bcl-2 homology 3 interacting-domain death agonist (Bid) (Mandal et al. 2008). As demonstrated in other cell types, Akt may promote murine double minute 2 (Mdm2)-mediated ubiquitination of p53 during β-selection (Ogawara et al. 2002). In support of Bim and Bid as executioners of apoptosis during β-selection, deletion of either Bim or Bid in pre-TCRα$^{-/-}$ mice significantly reduces the percentage of apoptotic DN3 thymocytes. However, in contrast to constitutive Akt activity, Bim or Bid deficiency does not allow further development to the DP stage (Hagenbeek et al. 2004; Mao et al. 2007; Mandal et al. 2008). Thus, downregulation of Bim and Bid are important for survival at the β-selection checkpoint, but additional signals from the pre-TCR are needed for differentiation and proliferation.

In addition to pre-TCR-mediated regulation of Bim and Bid through Akt activation, pre-TCR signaling also upregulates the anti-apoptotic protein Bcl2A1 but not Bcl-2, Mcl-1, or Bcl-$_{XL}$ (Mandal et al. 2005; Trampont et al. 2010). Pre-TCR induction of Bcl2A1 appears to be mediated by the protein kinase C (PKC) pathway and is independent of Akt (Mandal et al. 2005). Retroviral expression of Bcl2A1 in Rag1$^{-/-}$ thymocytes promotes their survival and differentiation *in vivo*, whereas knockdown of Bcl2A1 increases apoptosis of pre-TCR^{+} but not pre-TCR^{-} cell lines (Mandal et al. 2005). More recently, signaling through the chemokine receptor C-X-C motif receptor 4 (CXCR4) has been reported to provide co-stimulatory signals to the pre-TCR, converging on activation of extracellular signal-regulated kinase (ERK) 1/2 (Trampont et al. 2010). A role for ERK1/2 signaling in β-selection is established but its contribution to survival is not well understood (Michie and Zuniga-Pflucker 2002).

Deletion of CXCR4 impaired thymocyte survival during β-selection and this was at least due to decreased Bcl2A1 expression, implicating the ERK1/2 pathway as another regulator of its expression (Trampont et al. 2010).

The extrinsic apoptosis pathway initiated by death receptor signaling may also play a role at the β-selection checkpoint. Transgenic expression of a dominant-negative form of Fas-associated death domain (FADD), an adaptor molecule required for apoptosis induction downstream of multiple death receptors, partially rescued thymocyte development in $Rag1^{-/-}$ mice (Newton et al. 2000). Though Fas has been ruled out, it remains unclear which death receptors are involved.

Collectively, both inactivation of pro-apoptotic factors and induction of anti-apoptotic factors contribute to thymocyte survival during β-selection. Regulation of the Bcl-2 family by the PI3K/Akt, PKC, and ERK1/2 pathways are important mechanisms utilized by the pre-TCR to mediate survival of thymocytes at the β-selection checkpoint.

2.3. Death by neglect

Successful β-selection initiates rearrangement of the TCRα locus and differentiation into DP thymocytes (**Figure 1**). The number of recombination events correlates with the lifespan of the cell. Long-lived DP thymocytes exhaust the TCRα locus, thereby maximizing the chance of producing a functional TCR and engaging a positively selecting pMHC (Guo et al. 2002). Yet this is a relatively rare fate; most DP thymocytes undergo death by neglect upon failure to receive survival signals. This section discusses the factors that control DP survival in the absence of signaling through Notch1, IL-7 receptor, and the TCR (Huang et al. 2003; Yu et al. 2004). A prominent theory in the past was that glucocorticoids produced by thymic epithelial cells induce death by neglect in DP thymocytes that have not received TCR-induced resistance (Cohen 1992; Vacchio et al. 1994). However, studies in mice with hematopoietic-specific glucocorticoid receptor deficiency suggest that glucocorticoids do not have a significant role in death by neglect (Brewer et al. 2002; Purton et al. 2002).

Though it has its limitations, measuring spontaneous thymocyte death *in vitro* is a common way to study death by neglect. The extrinsic apoptosis pathway does not appear to play a significant role as blocking Fas/Fas ligand interactions or expression of dominant-negative FADD in thymocytes does not impair spontaneous death (Newton et al. 1998; Zhang et al. 2000). In contrast, many factors involved in the intrinsic apoptosis pathway are implicated in death by neglect. For example, DP thymocytes deficient for the pro-apoptotic proteins Bim or p53-upregulated modulator of apoptosis (Puma) have increased viability *in vitro* compared to wildtype DP, and deletion of both Bim and Puma further improves survival (Bouillet et al. 1999; Erlacher et al. 2006). In addition, $Bax^{-/-}$ $Bak^{-/-}$ thymocytes are resistant to spontaneous death (Rathmell et al. 2002). Of the anti-apoptotic Bcl-2 family members, numerous studies have shown that $Bcl\text{-}x_L$ plays a critical role in counteracting death by neglect. For one, the expression pattern of $Bcl\text{-}x_L$ strongly suggests that it promotes survival during the TCR-independent DP phase since spontaneous death over time correlates with decreased Bcl-x_L levels (Zhang et al. 2000), and Bcl-x_L expression is mostly restricted to DP thymocytes *in vivo* (Ma et al. 1995). Furthermore, deletion of the Bcl-x gene renders DP thymocytes more susceptible to apoptosis *in vitro* (Ma et al. 1995; Zhang and He 2005). Deletion of Mcl-1 also impairs DP thymocyte survival *in vitro* (Dzhagalov et al. 2008). While Mcl-1 deficiency results in a more modest loss of DP thymocytes *in vivo* compared to Bcl-x_L deficiency, combined deletion of Mcl-1 and Bcl-x_L results in a severe reduction in DP numbers, suggesting that both factors are important in preventing death by neglect.

The transcription factor retinoic acid-related orphan receptor (ROR) γt is a key activator of Bcl-x_L expression in DP cells. Similar to Bcl-x_L deficiency, deletion of RORγt results in increased DP apoptosis *in vivo* and reduced DP viability *in vitro* (Sun et al. 2000). Impaired DP survival in the absence of RORγt is likely due to reduced Bcl-x_L expression as complementation of RORγt$^{-/-}$ mice with transgenic Bcl-x_L rescues survival (Sun et al. 2000). Furthermore, decreased processivity of the TCRα locus is found in RORγt$^{-/-}$ mice, while the opposite is true for Bcl-x_L transgenic mice, highlighting the importance of these survival factors to the generation of a functional and self-restricted TCR (Guo et al. 2002). Pre-TCR signaling upregulates RORγt expression; however, RORγt activity is temporarily inhibited to allow generation of a large DP pool following β-selection. Active RORγt inhibits the cell cycle and promotes survival of resting DP thymocytes (Xi and Kersh 2004; Xi et al. 2006). Consistent with the importance of RORγt and Bcl-x_L in preventing death by neglect, multiple pathways have been reported to regulate their expression. For example, pre-TCR signaling induces expression of the transcription factor T cell factor 1 (TCF-1), which associates with β-catenin to execute Wnt signaling (Goux et al. 2005). TCF-1$^{-/-}$ thymocytes have reduced expression of RORγt and Bcl-x_L (Yuan et al. 2010); only TCF-1 isoforms that can interact with β-catenin are able to induce Bcl-x_L expression and rescue survival of TCF-1$^{-/-}$ DP thymocytes (Ioannidis et al. 2001). These data suggest that pre-TCR signaling and Wnt signaling cooperate to promote DP survival. Contrary to these reports, stabilization of β-catenin has been shown to impair DP survival (Gounari et al. 2001). Since β-catenin provides the transactivation domain and is thus the limiting factor in β-catenin/TCF-1-mediated transcription, a possible explanation is that physiological TCF-1 and β-catenin interactions promote survival while constitutively active signaling triggers tumour suppressors to induce apoptosis. The PI3K/Akt pathway reprises its role as a key mediator of thymocyte survival by also opposing death by neglect. Expression of constitutively active Akt enhances DP thymocyte survival in media and in fetal thymic organ cultures, while ablation of Akt or PI3K results in increased DP apoptosis (Jones et al. 2000; Swat et al. 2006; Mao et al. 2007). Specifically, the isoforms Akt1, Akt2, and PI3Kδ are implicated in DP survival, with the role of PI3Kγ$^{-/-}$ being more controversial (Sasaki et al. 2000; Swat et al. 2006; Mao et al. 2007). Bcl-x_L induction is at least part of the mechanism by which Akt promotes DP survival (Jones et al. 2000). Interestingly, Akt is a negative regulator of glycogen synthase kinase 3 (GSK3), a kinase that inhibits Wnt signaling through destabilization of β-catenin (Gounari et al. 2001; Song et al. 2005). Thus, in addition to upregulation of TCF-1, pre-TCR signaling may synergize with the Wnt pathway through Akt-mediated stabilization of β-catenin.

Promotion of glucose uptake and glycolysis by Akt in pre-β-selection DN3 thymocytes was previously mentioned (Ciofani and Zuniga-Pflucker 2005). Unlike DN thymocytes, only a small fraction of DP express glucose transporter 1 and it is unknown whether Akt regulates its expression (Swainson et al. 2005). However, conservation of energy and enhancement of energy production do appear to play an important role in extending the lifespan of DP thymocytes. It was recently reported that liver kinase B1, which activates ADP-activated protein kinase (AMPK) in response to ATP depletion, promotes DP survival *in vivo* and *in vitro* (Cao et al. 2010). Activated AMPK enacts metabolic changes to promote cell survival, and in DP thymocytes, its mechanism appears to include RORγt and Bcl-xL expression (Cao et al. 2010). Aside from TCF-1 and RORγt, the transcription factor c-myb is also thought to induce Bcl-xL expression in DP thymocytes in a TCF-1 and RORγt-independent manner (Yuan et al. 2010).

Taken together, these studies show that Mcl-1 and Bcl-xL play critical roles in preventing death by neglect of DP thymocytes, in part due to antagonism of pro-apoptotic Bcl-2 family members. Despite involvement of the intrinsic apoptosis pathway in death by neglect, components of the apoptosome, caspase-9 and apoptotic protease-activating factor 1 (Apaf-1), have been found to be mostly dispensable in spontaneous thymocyte death (Marsden et al. 2002). It was proposed that spontaneous death in the absence of caspase-9 and Apaf-1 is mediated by a low level of active caspase-7. However, another study found that z-DEVD, an inhibitor with preference for caspases-3 and -7, does not inhibit spontaneous thymocyte apoptosis (Zhang et al. 2000). Both studies reported that pan-caspase inhibitors such as z-VAD and IDN-1965 partially block spontaneous death. While pharmacological inhibitors have limitations, these data suggest the possible involvement of other caspases and/or caspase-independent cell death mechanisms.

3. TCR-dependent T cell development

The vast majority of DP thymocytes die by neglect due to failure of their TCRs to interact with self-pMHC; the fate of the remainder is determined by the affinity of this interaction. DP thymocytes that experience low affinity TCR stimulation undergo positive selection, receiving cues for survival, migration from the cortex to the medulla, and differentiation into CD4 or CD8 single positive (SP) thymocytes (**Figure 1**). In contrast, high affinity TCR-pMHC interactions result in negative selection, which is primarily mediated by clonal deletion of thymocytes expressing the high affinity TCR. Low affinity TCR ligands are thought to be non-cognate self-peptides, whereas cognate antigen provides high affinity stimulation (Starr et al. 2003). During migration through the thymic cortex, DP thymocytes may receive both low and high affinity TCR signals, but negative selection is dominant over positive selection. These processes are strictly controlled and dysregulation can lead to the development of immunodeficiency and autoimmune disorders. The remainder of this chapter discusses TCR-induced signaling pathways during negative and positive selection and subsequent regulation of pro- and anti-apoptotic factors.

3.1. TCR signaling pathways in positive and negative selection

An important unresolved question in T cell development is how high and low affinity TCR stimulation is translated into negative and positive selection outcomes. The current model is centered on differential activation of the mitogen-activated protein kinases (MAPKs) ERK1/2, ERK5, p38, and c-Jun N-terminal kinase (JNK) (**Figure 2**). MAPK signaling cascades, which involve activation of a series of kinases (MEKK→MEK→MAPK), mediate responses to extracellular stimuli including TCR stimulation. ERK1/2 is known to promote survival and differentiation while JNK and p38 are linked to apoptosis in other systems (Xia et al. 1995). Several lines of evidence suggest that they have similar functions during thymocyte development. For example, deletion of JNK1 or JNK2 renders thymocytes more resistant to death upon anti-CD3ε stimulation (Sabapathy et al. 1999; Sabapathy et al. 2001). Consistent with this, inhibition of JNK activity by a dominant-negative mutant inhibits peptide-induced negative selection *in vivo* (Rincon et al. 1998). Likewise, addition of a p38 inhibitor to a TCR transgenic fetal thymic organ culture impairs peptide-induced deletion (Sugawara et al. 1998). Characterization of the MEK5-ERK5 pathway is relatively recent compared to other MAPK signaling cascades. Dominant-negative ERK5 and MEK5 inhibit thymocyte apoptosis *in vitro* and peptide-induced deletion in some models of negative selection, respectively (Fujii et al. 2008; Sohn et al. 2008). While JNK, p38, and ERK5 have been implicated in negative selection, they are not required for positive selection of thymocytes (Rincon et al. 1998; Sugawara et al. 1998; Sohn et al. 2008). Conversely, inhibition of ERK1/2 or their activator MEK blocks positive selection but does not affect negative selection (Alberola-Ila et al. 1995; Alberola-Ila et al. 1996; Sugawara et al. 1998; Pages et al. 1999).

The upstream molecules that link high and low affinity TCR stimulation to differential MAPK activation are not well understood. Whereas premature TCRα expression inhibits β-selection, the TCRα chain is essential for positive selection of DP thymocytes (Mombaerts et al. 1992; Takahama et al. 1992; Lacorazza et al. 2001). This is because pMHC-induced signals impair development at the DN stage, whereas selection of DP thymocytes is dependent on pMHC ligands. It has been shown that CD3δ is required to transduce positive selection but not β-selection signals (Dave et al. 1997). Interestingly, the TCRα chain contains a motif important for both peptide contact and retention of CD3δ in the TCR complex, suggesting that CD3δ is a critical link between ligand binding and signal transduction (Backstrom et al. 1998). Mutating this motif in the TCRα chain abrogates CD3δ association with the TCR and ERK1/2 activation, resulting in defective peptide-induced positive selection (Werlen et al. 2000). This mutation has no effect on p38 and JNK activation or negative selection in the same system. During negative selection, JNK activation may be connected to the TCR complex through an upstream kinase called misshapen/NIKs-related kinase (MINK). This is evidenced by an association between CD3ε, MINK, and the adaptor protein non-catalytic region of tyrosine kinase (Nck) after stimulation of TCR transgenic thymocytes with cognate peptide (McCarty et al. 2005) (**Figure 2**). Consistent with its role in activating JNK, inhibition of MINK activity impairs negative selection *in vivo*. Involvement of different CD3 chains may result in differential phosphorylation of linker for activation of T cells (LAT), the central adaptor protein that links TCR proximal and distal signaling pathways (Starr et al. 2003). For example, regulation of different MAPK pathways through the adaptor protein growth factor receptor-bound protein 2 (Grb2) and the Ras activating protein Ras guanyl-releasing protein 1 (RasGRP1) in TCR signaling is thought to result from phosphorylation of different LAT residues (Wange 2000). Grb2 was shown to be important for p38 and JNK activation and negative selection, whereas RasGRP1 is essential for ERK1/2 activation and positive selection (Dower et al. 2000; Gong et al. 2001). However, a recent study reported that $Grb2^{-/-}$ mice are impaired in both negative and positive selection (Jang et al. 2010). Though much remains unknown about the discrimination of positive and negative selection signals, these findings shed light on how DP thymocyte fate is determined by TCR-pMHC interactions.

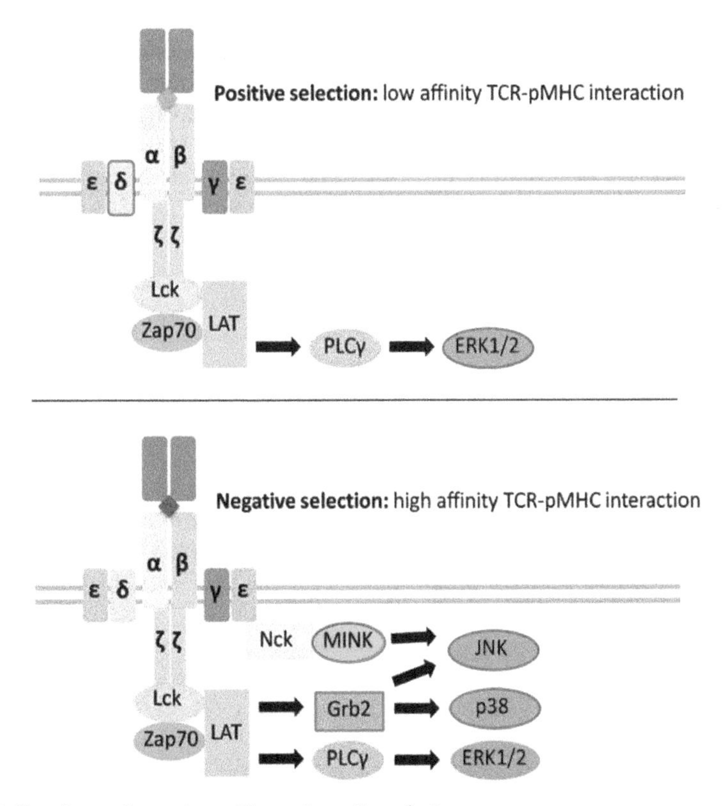

Figure 2. Signaling pathways in positive and negative selection.

Selection of DP thymocytes depends on the affinity of the TCR for self-pMHC in the thymus. Low affinity TCR-pMHC interactions result in positive selection and high affinity interactions in negative selection. The TCR is composed of one α and β chain each and transduces signals through a complex consisting of CD3 chains (γ, δ, ε, ζ). TCR proximal signaling events involve activation of kinases Lck and Zap70. Differential phosphorylation of the adaptor protein LAT is thought to result in activation of different MAPK pathways. The MAPKs JNK and p38 are important for negative selection and ERK1/2 for positive selection. JNK activation has also been linked to the TCR complex through the adaptor protein Nck and kinase MINK. ERK5 is another, relatively uncharacterized MAPK that may contribute to negative selection (not depicted). Active MAPK pathways are thought to lead to induction of pro-survival and pro-death factors that mediate positive or negative selection. (Lck - lymphocyte-specific protein tyrosine kinase; Zap70 - zeta chain-associated protein kinase 70.)

3.2. Induction of survival factors during positive selection

TCR signaling during positive selection results in reacquisition of IL-7 responsiveness in post-selection DP and SP thymocytes (Van De Wiele et al. 2004; Marino et al. 2010). Neutralizing IL-7Rα has been shown to inhibit SP development upon transfer of thymocytes from a non-selection (MHC$^{-/-}$) to a positive selection (MHC$^{+/+}$) background (Akashi et al. 1997). However, IL-7R$\alpha^{-/-}$ mice have a normal frequency of SP thymocytes (Peschon et al. 1994). Characterization of the role of IL-7 in positive selection may be confounded by the requirement for IL-7 in DN survival and proliferation. While IL-7 may not be required for positive selection, it is thought that IL-7 signaling provides important survival cues in SP thymocytes. Along with IL-7 responsiveness, Bcl-2 and Mcl-1 expression are upregulated in SP thymocytes (Linette et al. 1994; Akashi et al. 1997; Marino et al. 2010). Since Mcl-1 regulates early DN survival, a CD4-cre recombinase system was used to conditionally delete Mcl-1 at the DP stage. Positive selection is impaired in the absence of Mcl-1, as indicated by a reduced number of SP thymocytes (Dzhagalov et al. 2008; Dunkle et al. 2010). Mcl-1 deficiency is partially rescued by transgenic Bcl-2 expression, suggesting that Mcl-1 and Bcl-2 act on overlapping and distinct targets (Dunkle et al. 2010). Likewise, Bcl-2$^{-/-}$ mice exhibit a partial decrease in SP numbers, consistent with the idea of other proteins providing redundant and non-redundant functions during positive selection (Wojciechowski et al. 2007). Bcl-2$^{-/-}$ thymi are marked by a high frequency of DNA fragmentation, as indicated by positive staining in the terminal deoxynucleotidyl transferase-mediated dUTP nick end labeling (TUNEL) assay, as well as loss of thymocytes in the medulla, where SP cells normally reside (Veis et al. 1993). Conversely, transgenic Bcl-2 expression enhances positive selection and even allows SP development in the absence of MHC, though additional pMHC-induced signals are required for full maturation (Linette et al. 1994; Williams et al. 1998). The transcription factor c-Fos has been identified as an activator of Bcl-2 expression and also promotes positive selection on polyclonal and transgenic TCR backgrounds (Wang et al. 2009). Interestingly, c-Fos is implicated as a sensor for ERK1/2 signal duration (Murphy et al. 2002). Therefore, sustained ERK1/2 signaling during positive selection may be translated into c-Fos stabilization and Bcl-2 induction. Despite promoting thymocyte survival during positive selection, Bcl-2 is limited in its ability to inhibit clonal deletion of autoreactive thymocytes during negative selection (Sentman et al. 1991; Strasser et al. 1991).

3.3. The role of intrinsic and extrinsic apoptosis pathways in clonal deletion

Clonal deletion is widely held to occur through apoptosis, which can be mediated by extrinsic and intrinsic pathways. Many studies have examined the role of death receptors in negative selection. Mice defective for Fas or Fas ligand have been shown to have normal deletion of autoreactive thymocytes in multiple TCR transgenic models (Sidman et al. 1992; Singer and Abbas 1994; Sytwu et al. 1996). Similarly, tumour necrosis factor-related apoptosis-inducing ligand (TRAIL) has been shown to have no effect on peptide- or superantigen-induced clonal deletion (Simon et al. 2001; Cretney et al. 2003; Cretney et al. 2008). However, other studies report that Fas and TRAIL signaling do contribute to negative selection under some conditions (Kishimoto et al. 1998; Lamhamedi-Cherradi et al. 2003). Many of these studies utilize models of negative selection in which exogenous antigen is injected *in vivo*. This can cause non-specific thymocyte deletion by the secreted products of activated peripheral T cells (Martin and Bevan 1997). Because activation-induced cell death of peripheral T cells is impaired in the absence of death receptors, results derived from these model systems may be complicated by involvement of extrathymic factors (Singer and Abbas 1994; Sytwu et al. 1996). Importantly, transgenic expression of dominant-negative FADD or the viral caspase-8 inhibitor cytokine response modifier A (CrmA) does not impair negative selection, strongly suggesting that the extrinsic apoptosis pathway is dispensable (Smith et al. 1996; Walsh et al. 1998; Newton et al. 2000).

As discussed in detail in the following section, there is strong evidence that the intrinsic apoptosis pathway is involved in negative selection. Caspase-3 is widely held to be the main executioner caspase in mammals since both death receptor and mitochondrial-initiated pathways converge on its activation. Activation of caspase-3 is observed in thymocytes stimulated *in vitro* and from TCR transgenic models of negative selection (Alam et al. 1997; Hu et al. 2009). In contrast, the active forms of executioner caspases-6 and -7 are not detected after stimulation of TCR transgenic thymocytes with cognate peptide but can be induced by non-specific stimulation with staurosporine (Hara et al. 2002). Furthermore, comparison of caspase-3$^{-/-}$, caspase-7$^{-/-}$, and double knockout thymocytes indicates that caspase-3 is mainly responsible for DNA fragmentation in response to anti-CD3ε and anti-CD28 stimulation (Lakhani et al. 2006). Taken together, these data suggest that caspase-3 is the primary executioner caspase in clonal deletion. Though it normally plays a major role in clonal deletion, caspase-3 activation is not strictly required for negative selection *in vivo* (Hu et al. 2009; Murakami et al. 2010). While a role for other executioner caspases in negative selection has not been excluded, numerous studies using pan-caspase inhibitors *in vitro* and *in vivo* have shown that TCR-induced thymocyte death can occur in the absence of caspase activity, albeit to reduced levels in some systems (Alam et al. 1997; Izquierdo et al. 1999; Doerfler et al. 2000; Hara et al. 2002). These data highlight the existence of multiple mechanisms that have evolved to mediate negative selection and self-tolerance.

3.4. Initiators of the intrinsic apoptosis pathway

Mitochondria are central to the initiation of caspase-dependent and caspase-independent cell death (Jaattela and Tschopp 2003). Of the Bcl-2 family that controls the mitochondrial gateway to cell death, the pro-apoptotic protein Bim has a critical role in TCR-induced thymocyte death. Bim induction has been demonstrated after TCR stimulation of thymocytes *in vitro* and in numerous models of negative selection *in vivo* (Bouillet et al. 2002; Schmitz et al. 2003; Huang et al. 2004; Zucchelli et al. 2005; Baldwin and Hogquist 2007; Liston et al. 2007). The JNK pathway positively regulates Bim expression and function in other cell types (Whitfield et al. 2001; Putcha et al. 2003). Post-translational modification does not seem to be a major mechanism by which the TCR regulates Bim in thymocytes (Bunin et al. 2005). In thymocytes, MINK activity is linked to JNK activation and Bim induction, though it was not shown that Bim induction is JNK-dependent (McCarty et al. 2005). Another study found that PKC inhibitors, but not JNK, p38, or MEK inhibitors, block Bim induction in thymocytes stimulated with anti-CD3ε and anti-CD28 (Cante-Barrett et al. 2006). However, unlike JNK and p38, PKC activity is not required for negative selection (Anderson et al. 1995; Sun et al. 2000). More work is required to clarify the pathways activated during negative selection that lead to regulation of Bim.

Bim is thought to have a critical role in clonal deletion as Bim deficiency results in resistance to TCR-induced thymocyte apoptosis *in vitro* (Bouillet et al. 2002) and abrogation of caspase-3 activation in DP thymocytes undergoing negative selection *in vivo* (Hu et al. 2009). Consistent with its essential role in thymocyte apoptosis, deletion of DP thymocytes by superantigen and cognate peptide is impaired in the absence of Bim (Bouillet et al. 2002). Though deletion of DP thymocytes is indicative of clonal deletion, the number of autoreactive SP thymocytes is the most accurate measure of negative selection. In the physiological HY^{cd4} model of negative selection, Bim deficiency delays deletion of DP thymocytes but the number of autoreactive SP thymocytes is ultimately comparable in the presence and absence of Bim (Hu et al. 2009). Because Bim is required for caspase-3 activation in this model, these data support the idea of a redundant non-apoptotic cell death mechanism of negative selection. In addition, superantigen-mediated negative selection, as measured by autoreactive SP numbers, has also been reported to be Bim-independent (Jorgensen et al. 2007).

Another factor to consider when evaluating the role of a protein in negative selection is the model system utilized. The molecular events involved in superantigen-induced deletion may be different from those initiated from pMHC interactions. Mice expressing transgenic TCRs against endogenous or neo self-antigens (thus avoiding the issues associated with peripheral T cell activation) have been the most powerful tools available for studying negative selection. However, there are key differences between TCR transgenic models. For example, in the classical HY model where HY is a male-specific antigen, the transgenic HY TCR is prematurely expressed on DN thymocytes such that deletion occurs during the DN to DP transition (Takahama et al. 1992; Baldwin et al. 2005). Thus, while Bim deficiency impairs negative selection in the HY model, this may in part reflect a role for Bim in DN thymocyte death (Bouillet et al. 2002). The HY^{cd4} model utilizes a CD4-cre recombinase system to conditionally express the transgenic HY TCRα chain at the DP stage, allowing selection to occur during the DP to SP transition as in wildtype mice (Baldwin et al. 2005). When the timing of TCR expression has been corrected, Bim appears to be dispensable for deletion of HY TCR^+ thymocytes (Hu et al. 2009; Kovalovsky et al. 2010).

The stage of development at which deletion occurs has considerable implications on the molecular mechanism involved due to each thymocyte subset having differential gene expression, signaling threshold, and localization. For example, positive selection induces differentiation of DP into SP and migration to the medulla. Localization affects antigen presentation because the thymic cortex and medulla contain different types of antigen presenting cells. Importantly, ectopic expression of tissue-restricted antigens such as insulin is restricted to medullary thymic epithelial cells (Derbinski et al. 2001; Anderson et al. 2002). Because HY is a ubiquitous antigen, clonal deletion of HY^{cd4} thymocytes occurs at the DP stage in the cortex, a process that does not require Bim (McCaughtry et al. 2008; Hu et al. 2009). While past studies have cited defective Bim induction and clonal deletion in type I diabetes-prone non-obese diabetic (NOD) mice (Zucchelli et al. 2005; Liston et al. 2007), a recent study clarified that NOD mice do not have a cell-intrinsic impairment in clonal deletion (Mingueneau et al. 2012). Negative selection against tissue-restricted antigens was also recently examined by transfer of OT-I or OT-I $Bim^{-/-}$ TCR transgenic bone marrow, which specifically recognizes ovalbumin peptide, into recipients in which the cognate antigen is driven by the ubiquitously active actin promoter or the tissue-restricted rat insulin promoter (Suen and Baldwin 2012). In agreement with results from the HY^{cd4} model, Bim is not required for negative selection against ubiquitous antigen, but is required for negative selection against tissue-restricted antigen in a cell-intrinsic way (**Table 1**). Because DP thymocytes are more sensitive to TCR-induced death than SP thymocytes (Davey et al. 1998), one explanation is that either a Bim-dependent or independent mechanism is sufficient to kill DP cells, while both are required for SP deletion. Alternatively, interactions with medullary thymic epithelial cells may not induce factors that mediate Bim-independent cell death.

The Bcl-2 family are critical regulators of mitochondrial integrity. The balance of pro-apoptotic and anti-apoptotic members controls the activation of Bax and Bak, which leads to cytochrome c release and caspase activation. Programmed cell death plays an important role in the elimination of dysfunctional and autoreactive thymocytes. This table summarizes some of the proteins known to play a role in thymocyte survival. Different Bcl-2 members are important at different stages of thymocyte development.

One candidate for mediating Bim-independent clonal deletion is the NR4A nuclear receptor Nur77. The NR4A nuclear receptor family is comprised of three proteins closely related in structure and function: Nur77, Nor-1, and Nurr1, though only Nur77 and Nor-1 are induced in stimulated thymocytes (Cheng et al. 1997). Nur77 is induced by TCR stimulation of thymocytes and DO11.10 T cell hybridoma cells *in vitro* (Liu et al. 1994; Woronicz et al. 1994) and is consistently among the list of genes upregulated during negative selection *in vivo* (Schmitz et al. 2003; Huang et al. 2004; Zucchelli et al. 2005; Baldwin and Hogquist 2007; Liston et al. 2007). Deletion of the transactivation domain of Nur77 creates a dominant-negative mutant that interferes with the transcriptional activity of all NR4A family members (Cheng et al. 1997). Expression of dominant-negative Nur77 partially impairs clonal deletion in some TCR transgenic models but not others (Calnan et al. 1995; Zhou et al. 1996). Past studies with $Nur77^{-/-}$ mice reported normal negative selection *in vivo*, suggesting that Nor-1 provides redundant functions in thymocytes (Lee et al. 1995). Nevertheless, little is known about Nor-1 compared to Nur77. Recently, Nur77 deficiency alone has also been shown to impair negative selection (Fassett et al. 2012). The MEK5-ERK5 pathway has been reported to induce Nur77 expression in thymocytes and DO11.10 cells (Kasler et al. 2000; Sohn et al. 2008). However, MEK5 and ERK5 have not been shown to be required for Nur77 induction in response to TCR stimulation. Indeed, ERK5 is not necessary for Nur77 induction upon activation of peripheral T cells (Ananieva et al. 2008). Consistent with dominant-negative Nur77 studies, expression of dominant-negative MEK5 inhibits clonal deletion in certain TCR transgenic models (Sohn et al. 2008). Dominant-negative ERK5 has been shown to inhibit apoptosis in DO11.10 cells *in vitro* but characterization of ERK5 in negative selection *in vivo* is presently limited. Because pharmacological inhibitors can act on both MEK1/2 (upstream of ERK1/2) and MEK5 at high doses (Mody et al. 2001), it is possible that MEK5 can act on additional targets other than ERK5. Taken together, these data support a role for Nur77 in some types of negative selection.

	Pre-β-selection	β-selection	Death by neglect	TCR-dependent selection
Anti-apoptotic members				
Bcl-2	✔			✔
Mcl-1	✔		✔	✔
Bcl2A1		✔		
Bcl-x_L			✔	
Pro-apoptotic members				
Bad	✔			
Bim		✔	✔	✔
Bid		✔		
Puma			✔	

Table 1. Regulation of survival by the Bcl-2 family during thymocyte development.

The mechanism of Nur77-induced thymocyte death is controversial. Nur77 was initially thought to upregulate factors that mediate apoptosis since the transcriptional activity of Nur77 correlates with its ability to induce thymocyte death (Kuang et al. 1999). Furthermore, Nur77 has been shown to remain in the nucleus of stimulated thymocytes, while nuclear export of Nur77 in mature T cells is thought to protect them from apoptosis (Cunningham et al. 2006). However, the only target genes of Nur77 with known apoptotic function are those involved in the extrinsic apoptosis pathway (Rajpal et al. 2003). Thus, the role of Nur77-mediated transcription in negative selection has been questioned. Studies of Nur77 function in cancer cells have identified a mitochondrial, transcription-independent cell death mechanism whereby Nur77 converts Bcl-2 into a pro-apoptotic protein by exposing its Bcl-2 homology 3 (BH3) domain (Lin et al. 2004). By utilizing a subcellular fractionation protocol not used in previous studies, Nur77 was reported to translocate to mitochondria and associate with Bcl-2 upon TCR stimulation of thymocytes (Thompson and Winoto 2008). While other groups also report mitochondrial translocation of Nur77 following stimulation (Stasik et al. 2007; Wang et al. 2009), whether Nur77 mediates thymocyte death through Bcl-2 conversion is controversial. In stimulated DO11.10 cells, Bcl-2 and Nur77 do not interact and Bcl-2 expression protects against Nur77-mediated death (Wang et al. 2009). Furthermore, Bcl-XL, not Bcl-2, is the predominant survival factor expressed in DP thymocytes; transgenic overexpression of Bcl-2 was required to detect interaction with Nur77 (Ma et al. 1995; Thompson and Winoto 2008). Studies also differ on regulation of Nur77 nuclear export. In DO11.10 cells, phosphorylation of serine 354 by the ERK1/2-ribosomal S6 kinase (RSK) pathway is necessary for nuclear export (Wang et al. 2009). This is contested by another study that found PKC but not ERK1/2 to be required for mitochondrial translocation (Thompson et al. 2010). Since ERK1/2 (Alberola-Ila et al. 1995; Alberola-Ila et al. 1996; Sugawara et al. 1998; Pages et al. 1999) and PKC (Anderson et al. 1995; Sun et al. 2000) are not required for negative selection, the contribution of the transcription-independent mechanism of Nur77 to clonal deletion is also questionable.

Mice that express transgenic Nur77 or transgenic Nor-1 have severely reduced thymic cellularity and an increased frequency of TUNEL$^+$ thymocytes (Calnan et al. 1995; Cheng et al. 1997). Conversely, thymocytes from DN-Nur77 mice are more resistant to DNA fragmentation following anti-CD3ε stimulation (Zhou et al. 1996). Surprisingly, transgenic Nur77 expression does not appear to be sufficient to induce cytochrome c release (Rajpal et al. 2003). A caveat of using these transgenic mice to study negative selection is that Nur77 transgene expression is driven by promoters that are active in DN thymocytes independently of TCR signaling. Thus, transgenic Nur77 may induce thymocyte death through transcription of extrinsic apoptosis genes, while physiological regulation of Nur77 by TCR signaling may favor a transcription-independent mechanism. Studies using T cell hybridomas generally support activation of the intrinsic apoptosis pathway by Nur77. For example, nuclear export of Nur77 in DO11.10 cells leads to cytochrome c release, caspase-9 cleavage, and poly(ADP-ribose) polymerase (PARP) cleavage (Wang et al. 2009). Though T cell hybridomas are more amenable to manipulation than thymocytes, it is important to keep in mind that the mechanism of Nur77-mediated death may differ between cell types. Despite induction of apoptosis, z-VAD treatment and Bcl-2 or Bcl-XL expression only partially rescues Nur77-induced cell death, suggesting contribution of a caspase-independent mechanism. Interestingly, Nur77 has been shown to mediate caspase-independent death in other cell types (Kim et al. 2003; Castro-Obregon et al. 2004; Lucattelli et al. 2006). Furthermore, it is unknown whether the DNA fragmentation induced by Nur77 in thymocytes is caspase-dependent oligonucleosomal fragmentation or caspase-independent large scale DNA fragmentation since the TUNEL assay does not discriminate between the two types (Ribeiro et al. 2006).

Abbreviation	Full name	Description
AIF	Apoptosis-inducing factor	Protein released from mitochondria that mediates cell death
AMPK	ADP-activated protein kinase	Regulates cellular energy metabolism in response to ATP depletion
Apaf-1	Apoptotic protease-activating factor 1	Activates caspase-9 when bound to cytochrome c and dATP
Bcl-2	B cell lymphoma 2	Founding, anti-apoptotic member of the Bcl-2 family of proteins, initially described in B cell lymphomas. Bcl-xL and Bcl2A1 are similarly named.
Bad	Bcl-2-associated death promoter	Pro-apoptotic member of the Bcl-2 family
Bak	Bcl-2 homologous antagonist/killer	Pro-apoptotic member of the Bcl-2 family that forms channels in the mitochondrial outer membrane
Bax	Bcl-2-associated X protein	Pro-apoptotic member of the Bcl-2 family that forms channels in the mitochondrial outer membrane
Bid	Bcl-2 homology 3 interacting-domain death agonist	Pro-apoptotic member of the Bcl-2 family
Bim	Bcl-2-interacting mediator of cell death	Pro-apoptotic member of the Bcl-2 family
CXCR4	C-X-C motif receptor 4	Chemokine receptor
DN	Double negative	$CD4^-CD8^-$ thymocyte subset
DP	Double positive	$CD4^+CD8^+$ thymocyte subset
ERK	Extracellular signal-regulated kinase	A subfamily of mitogen-activated protein kinases
FADD	Fas-associated death domain	Adaptor protein involved in signaling through death receptors
Grb2	Growth factor receptor-bound protein 2	Adaptor protein involved in T cell receptor signaling
JNK	c-Jun N-terminal kinase	A subfamily of mitogen-activated protein kinases
MAPK	Mitogen activated protein kinase	A class of kinases that respond to extracellular stimuli
Mcl-1	Myeloid leukemia sequence 1	Anti-apoptotic member of the Bcl-2 family
MHC	Major histocompatibility complex	Protein that presents peptide antigen to T cell receptor

MINK	Misshapen/NIKs-related kinase	Kinase involved in T cell receptor signaling that promotes JNK activation
Nck	Non-catalytic region of tyrosine kinase	Adaptor protein involved in T cell receptor signaling
NOD	Non-obese diabetic	Mouse strain genetically predisposed to developing type I diabetes
PI3K	Phosphoinositide 3-kinase	Phosphorylates phosphatidylinositol (4,5)-biphosphate to generate phosphatidylinositol (3,4,5)-triphosphate, leading to Akt activation
PKC	Protein kinase C	Kinase involved in T cell receptor signaling
pMHC	Peptide-MHC	Peptide presented on an MHC molecule
PTEN	Phosphatase and tensin homolog	Dephosphorylates phosphatidylinositol (3,4,5)-triphosphate to generate phosphatidylinositol (4,5)-biphosphate, negatively regulating Akt activation
Puma	p53-upregulated modulator of apoptosis	Pro-apoptotic member of the Bcl-2 family
Rag	Recombination activating gene	Enzyme that mediates genetic recombination of T cell receptor loci
RasGRP1	Ras-guanyl releasing protein 1	Activates Ras through the exchange of bound GDP for GTP
RBP-J	Recombination signal binding protein for immunoglobulin kappa J region	Transcriptional regulator that activates transcription when bound to intracellular domain of Notch proteins
RORγt	Retinoic acid-related orphan receptor γt	Transcription factor that induces Bcl-$_{XL}$ expression in thymocytes
SP	Single positive	$CD4^+CD8^-$ or $CD4^-CD8^+$ thymocyte subsets
TCR	T cell receptor	Antigen receptor expressed by T cells
TRAIL	Tumour necrosis factor-related apoptosis-inducing ligand	Protein that induces apoptosis by binding to death receptors
TUNEL	Terminal deoxynucleotidyl transferase-mediated dUTP nick end labeling	Method of detecting DNA fragmentation by labeling the terminal end of fragments

Table 2. Abbreviations used multiple times throughout this chapter.

The molecular mechanisms of negative selection remain unclear despite intense investigation into the matter. Though Bim and Nur77 are implicated as key mediators of TCR-induced thymocyte death, many questions surround their role in physiological negative selection. Consideration of the model system and readout to be used will greatly assist future endeavors to clarify the molecular mechanisms of negative selection.

4. Conclusion

The highly regulated struggle between survival and death during thymocyte development underscores the need to generate functional yet self-tolerant T cells. At multiple checkpoints during development in the thymus, the balance must be tipped in favour of pro-death factors to allow elimination of dysfunctional and autoreactive cells. It is becoming increasingly apparent that programmed cell death is not restricted to apoptosis but also involves caspase-independent processes such as autophagy and necroptosis. While non-apoptotic programmed cell death is better characterized in peripheral T cells (Jaattela and Tschopp 2003), many studies have provided evidence that caspase-independent death can occur at multiple stages of thymocyte development. Two proteins implicated in caspase-independent death are apoptosis-inducing factor (AIF) and endonuclease G. Both have been shown to mediate cell death following translocation from mitochondria to the nucleus, where they execute caspase-independent DNA fragmentation (Jaattela and Tschopp 2003). It will be of great interest to determine if AIF and endonuclease G contribute to caspase-independent thymocyte death, and if their translocation is induced by Nur77, which has been linked to mitochondrial localization and caspase-independent cell death. One thing is clear: mitochondria are a key gateway to cell death. The Bcl-2 family of proteins control mitochondrial integrity through regulating formation of Bax/Bak channels and opening of the mitochondrial permeability transition pore (Tsujimoto and Shimizu 2000). Therefore, they may be poised to control different mechanisms of programmed cell death. Though genetic manipulation of mice has accelerated our understanding of thymocyte survival and death, much remains to be characterized about the molecular mechanisms involved. These complex, interwoven, and tightly regulated mechanisms are necessary to balance the need for a competent and self-tolerant immune system.

Author details

Qian Nancy Hu and Troy A. Baldwin
University of Alberta, Canada

5. References

Akashi, K., M. Kondo, et al. (1997). "Bcl-2 rescues T lymphopoiesis in interleukin-7 receptor-deficient mice." Cell 89(7): 1033-1041.

Alam, A., M. Y. Braun, et al. (1997). "Specific activation of the cysteine protease CPP32 during the negative selection of T cells in the thymus." J Exp Med 186(9): 1503-1512.

Alberola-Ila, J., K. A. Forbush, et al. (1995). "Selective requirement for MAP kinase activation in thymocyte differentiation." Nature 373(6515): 620-623.
Alberola-Ila, J., K. A. Hogquist, et al. (1996). "Positive and negative selection invoke distinct signaling pathways." J Exp Med 184(1): 9-18.
Ananieva, O., A. Macdonald, et al. (2008). "ERK5 regulation in naive T-cell activation and survival." Eur J Immunol 38(9): 2534-2547.
Anderson, G., K. L. Anderson, et al. (1995). "Intracellular signaling events during positive and negative selection of CD4+CD8+ thymocytes in vitro." J Immunol 154(8): 3636-3643.
Anderson, M. S., E. S. Venanzi, et al. (2002). "Projection of an immunological self shadow within the thymus by the aire protein." Science 298(5597): 1395-1401.
Arstila, T. P., A. Casrouge, et al. (1999). "A direct estimate of the human alphabeta T cell receptor diversity." Science 286(5441): 958-961.
Backstrom, B. T., U. Muller, et al. (1998). "Positive selection through a motif in the alphabeta T cell receptor." Science 281(5378): 835-838.
Baldwin, T. A. and K. A. Hogquist (2007). "Transcriptional analysis of clonal deletion in vivo." J Immunol 179(2): 837-844.
Baldwin, T. A., M. M. Sandau, et al. (2005). "The timing of TCR alpha expression critically influences T cell development and selection." J Exp Med 202(1): 111-121.
Bouillet, P., D. Metcalf, et al. (1999). "Proapoptotic Bcl-2 relative Bim required for certain apoptotic responses, leukocyte homeostasis, and to preclude autoimmunity." Science 286(5445): 1735-1738.
Bouillet, P., J. F. Purton, et al. (2002). "BH3-only Bcl-2 family member Bim is required for apoptosis of autoreactive thymocytes." Nature 415(6874): 922-926.
Brewer, J. A., O. Kanagawa, et al. (2002). "Thymocyte apoptosis induced by T cell activation is mediated by glucocorticoids in vivo." J Immunol 169(4): 1837-1843.
Bunin, A., F. W. Khwaja, et al. (2005). "Regulation of Bim by TCR signals in CD4/CD8 double-positive thymocytes." J Immunol 175(3): 1532-1539.
Calnan, B. J., S. Szychowski, et al. (1995). "A role for the orphan steroid receptor Nur77 in apoptosis accompanying antigen-induced negative selection." Immunity 3(3): 273-282.
Cante-Barrett, K., E. M. Gallo, et al. (2006). "Thymocyte negative selection is mediated by protein kinase C- and Ca2+-dependent transcriptional induction of bim [corrected]." J Immunol 176(4): 2299-2306.
Cao, Y., H. Li, et al. (2010). "The serine/threonine kinase LKB1 controls thymocyte survival through regulation of AMPK activation and Bcl-XL expression." Cell Res 20(1): 99-108.
Castro-Obregon, S., R. V. Rao, et al. (2004). "Alternative, nonapoptotic programmed cell death: mediation by arrestin 2, ERK2, and Nur77." J Biol Chem 279(17): 17543-17553.
Cheng, L. E., F. K. Chan, et al. (1997). "Functional redundancy of the Nur77 and Nor-1 orphan steroid receptors in T-cell apoptosis." EMBO J 16(8): 1865-1875.
Ciofani, M., T. M. Schmitt, et al. (2004). "Obligatory role for cooperative signaling by pre-TCR and Notch during thymocyte differentiation." J Immunol 172(9): 5230-5239.
Ciofani, M. and J. C. Zuniga-Pflucker (2005). "Notch promotes survival of pre-T cells at the beta-selection checkpoint by regulating cellular metabolism." Nat Immunol 6(9): 881-888.

Cohen, J. J. (1992). "Glucocorticoid-induced apoptosis in the thymus." Semin Immunol 4(6): 363-369.

Cretney, E., A. P. Uldrich, et al. (2003). "Normal thymocyte negative selection in TRAIL-deficient mice." J Exp Med 198(3): 491-496.

Cretney, E., A. P. Uldrich, et al. (2008). "No requirement for TRAIL in intrathymic negative selection." Int Immunol 20(2): 267-276.

Cunningham, N. R., S. C. Artim, et al. (2006). "Immature CD4+CD8+ thymocytes and mature T cells regulate Nur77 distinctly in response to TCR stimulation." J Immunol 177(10): 6660-6666.

Dave, V. P., Z. Cao, et al. (1997). "CD3 delta deficiency arrests development of the alpha beta but not the gamma delta T cell lineage." EMBO J 16(6): 1360-1370.

Davey, G. M., S. L. Schober, et al. (1998). "Preselection thymocytes are more sensitive to T cell receptor stimulation than mature T cells." J Exp Med 188(10): 1867-1874.

Derbinski, J., A. Schulte, et al. (2001). "Promiscuous gene expression in medullary thymic epithelial cells mirrors the peripheral self." Nat Immunol 2(11): 1032-1039.

Doerfler, P., K. A. Forbush, et al. (2000). "Caspase enzyme activity is not essential for apoptosis during thymocyte development." J Immunol 164(8): 4071-4079.

Dower, N. A., S. L. Stang, et al. (2000). "RasGRP is essential for mouse thymocyte differentiation and TCR signaling." Nat Immunol 1(4): 317-321.

Dunkle, A., I. Dzhagalov, et al. (2010). "Mcl-1 promotes survival of thymocytes by inhibition of Bak in a pathway separate from Bcl-2." Cell Death Differ 17(6): 994-1002.

Dzhagalov, I., A. Dunkle, et al. (2008). "The anti-apoptotic Bcl-2 family member Mcl-1 promotes T lymphocyte survival at multiple stages." J Immunol 181(1): 521-528.

Egerton, M., R. Scollay, et al. (1990). "Kinetics of mature T-cell development in the thymus." Proc Natl Acad Sci U S A 87(7): 2579-2582.

Engel, I. and C. Murre (1999). "Ectopic expression of E47 or E12 promotes the death of E2A-deficient lymphomas." Proc Natl Acad Sci U S A 96(3): 996-1001.

Erlacher, M., V. Labi, et al. (2006). "Puma cooperates with Bim, the rate-limiting BH3-only protein in cell death during lymphocyte development, in apoptosis induction." J Exp Med 203(13): 2939-2951.

Fassett, M. S., W. Jiang, et al. (2012). "Nuclear receptor Nr4a1 modulates both regulatory T-cell (Treg) differentiation and clonal deletion." Proc Natl Acad Sci U S A 109(10): 3891-3896.

Fujii, Y., S. Matsuda, et al. (2008). "ERK5 is involved in TCR-induced apoptosis through the modification of Nur77." Genes Cells 13(5): 411-419.

Godfrey, D. I., J. Kennedy, et al. (1993). "A developmental pathway involving four phenotypically and functionally distinct subsets of CD3-CD4-CD8- triple-negative adult mouse thymocytes defined by CD44 and CD25 expression." J Immunol 150(10): 4244-4252.

Gong, Q., A. M. Cheng, et al. (2001). "Disruption of T cell signaling networks and development by Grb2 haploid insufficiency." Nat Immunol 2(1): 29-36.

Gounari, F., I. Aifantis, et al. (2001). "Somatic activation of beta-catenin bypasses pre-TCR signaling and TCR selection in thymocyte development." Nat Immunol 2(9): 863-869.

Goux, D., J. D. Coudert, et al. (2005). "Cooperating pre-T-cell receptor and TCF-1-dependent signals ensure thymocyte survival." Blood 106(5): 1726-1733.
Guo, J., A. Hawwari, et al. (2002). "Regulation of the TCRalpha repertoire by the survival window of CD4(+)CD8(+) thymocytes." Nat Immunol 3(5): 469-476.
Hagenbeek, T. J., M. Naspetti, et al. (2004). "The loss of PTEN allows TCR alphabeta lineage thymocytes to bypass IL-7 and Pre-TCR-mediated signaling." J Exp Med 200(7): 883-894.
Haks, M. C., P. Krimpenfort, et al. (1999). "Pre-TCR signaling and inactivation of p53 induces crucial cell survival pathways in pre-T cells." Immunity 11(1): 91-101.
Han, H., K. Tanigaki, et al. (2002). "Inducible gene knockout of transcription factor recombination signal binding protein-J reveals its essential role in T versus B lineage decision." Int Immunol 14(6): 637-645.
Hara, H., A. Takeda, et al. (2002). "The apoptotic protease-activating factor 1-mediated pathway of apoptosis is dispensable for negative selection of thymocytes." J Immunol 168(5): 2288-2295.
Hernandez-Munain, C., B. P. Sleckman, et al. (1999). "A developmental switch from TCR delta enhancer to TCR alpha enhancer function during thymocyte maturation." Immunity 10(6): 723-733.
Hozumi, K., C. Mailhos, et al. (2008). "Delta-like 4 is indispensable in thymic environment specific for T cell development." J Exp Med 205(11): 2507-2513.
Hozumi, K., N. Negishi, et al. (2004). "Delta-like 1 is necessary for the generation of marginal zone B cells but not T cells in vivo." Nat Immunol 5(6): 638-644.
Hu, Q., A. Sader, et al. (2009). "Bim-mediated apoptosis is not necessary for thymic negative selection to ubiquitous self-antigens." J Immunol 183(12): 7761-7767.
Huang, E. Y., A. M. Gallegos, et al. (2003). "Surface expression of Notch1 on thymocytes: correlation with the double-negative to double-positive transition." J Immunol 171(5): 2296-2304.
Huang, Y. H., D. Li, et al. (2004). "Distinct transcriptional programs in thymocytes responding to T cell receptor, Notch, and positive selection signals." Proc Natl Acad Sci U S A 101(14): 4936-4941.
Ioannidis, V., F. Beermann, et al. (2001). "The beta-catenin--TCF-1 pathway ensures CD4(+)CD8(+) thymocyte survival." Nat Immunol 2(8): 691-697.
Irving, B. A., F. W. Alt, et al. (1998). "Thymocyte development in the absence of pre-T cell receptor extracellular immunoglobulin domains." Science 280(5365): 905-908.
Izquierdo, M., A. Grandien, et al. (1999). "Blocked negative selection of developing T cells in mice expressing the baculovirus p35 caspase inhibitor." EMBO J 18(1): 156-166.
Jaattela, M. and J. Tschopp (2003). "Caspase-independent cell death in T lymphocytes." Nat Immunol 4(5): 416-423.
Jang, I. K., J. Zhang, et al. (2010). "Grb2 functions at the top of the T-cell antigen receptor-induced tyrosine kinase cascade to control thymic selection." Proc Natl Acad Sci U S A 107(23): 10620-10625.
Jiang, D., M. J. Lenardo, et al. (1996). "p53 prevents maturation to the CD4+CD8+ stage of thymocyte differentiation in the absence of T cell receptor rearrangement." J Exp Med 183(4): 1923-1928.

Jones, R. G., M. Parsons, et al. (2000). "Protein kinase B regulates T lymphocyte survival, nuclear factor kappaB activation, and Bcl-X(L) levels in vivo." J Exp Med 191(10): 1721-1734.

Jorgensen, T. N., A. McKee, et al. (2007). "Bim and Bcl-2 mutually affect the expression of the other in T cells." J Immunol 179(6): 3417-3424.

Juntilla, M. M., J. A. Wofford, et al. (2007). "Akt1 and Akt2 are required for alphabeta thymocyte survival and differentiation." Proc Natl Acad Sci U S A 104(29): 12105-12110.

Kasler, H. G., J. Victoria, et al. (2000). "ERK5 is a novel type of mitogen-activated protein kinase containing a transcriptional activation domain." Mol Cell Biol 20(22): 8382-8389.

Khaled, A. R., K. Kim, et al. (1999). "Withdrawal of IL-7 induces Bax translocation from cytosol to mitochondria through a rise in intracellular pH." Proc Natl Acad Sci U S A 96(25): 14476-14481.

Khaled, A. R., W. Q. Li, et al. (2002). "Bax deficiency partially corrects interleukin-7 receptor alpha deficiency." Immunity 17(5): 561-573.

Kim, K., C. K. Lee, et al. (1998). "The trophic action of IL-7 on pro-T cells: inhibition of apoptosis of pro-T1, -T2, and -T3 cells correlates with Bcl-2 and Bax levels and is independent of Fas and p53 pathways." J Immunol 160(12): 5735-5741.

Kim, S. O., K. Ono, et al. (2003). "Orphan nuclear receptor Nur77 is involved in caspase-independent macrophage cell death." J Exp Med 197(11): 1441-1452.

Kishimoto, H., C. D. Surh, et al. (1998). "A role for Fas in negative selection of thymocytes in vivo." J Exp Med 187(9): 1427-1438.

Koch, U., E. Fiorini, et al. (2008). "Delta-like 4 is the essential, nonredundant ligand for Notch1 during thymic T cell lineage commitment." J Exp Med 205(11): 2515-2523.

Kondo, M., K. Akashi, et al. (1997). "Bcl-2 rescues T lymphopoiesis, but not B or NK cell development, in common gamma chain-deficient mice." Immunity 7(1): 155-162.

Kovalovsky, D., M. Pezzano, et al. (2010). "A novel TCR transgenic model reveals that negative selection involves an immediate, Bim-dependent pathway and a delayed, Bim-independent pathway." PLoS One 5(1): e8675.

Kuang, A. A., D. Cado, et al. (1999). "Nur77 transcription activity correlates with its apoptotic function in vivo." Eur J Immunol 29(11): 3722-3728.

Lacorazza, H. D., C. Tucek-Szabo, et al. (2001). "Premature TCR alpha beta expression and signaling in early thymocytes impair thymocyte expansion and partially block their development." J Immunol 166(5): 3184-3193.

Lakhani, S. A., A. Masud, et al. (2006). "Caspases 3 and 7: key mediators of mitochondrial events of apoptosis." Science 311(5762): 847-851.

Lamhamedi-Cherradi, S. E., S. J. Zheng, et al. (2003). "Defective thymocyte apoptosis and accelerated autoimmune diseases in TRAIL-/- mice." Nat Immunol 4(3): 255-260.

Lee, S. L., R. L. Wesselschmidt, et al. (1995). "Unimpaired thymic and peripheral T cell death in mice lacking the nuclear receptor NGFI-B (Nur77)." Science 269(5223): 532-535.

Li, W. Q., Q. Jiang, et al. (2004). "Interleukin-7 inactivates the pro-apoptotic protein Bad promoting T cell survival." J Biol Chem 279(28): 29160-29166.

Lin, B., S. K. Kolluri, et al. (2004). "Conversion of Bcl-2 from protector to killer by interaction with nuclear orphan receptor Nur77/TR3." Cell 116(4): 527-540.

Linette, G. P., M. J. Grusby, et al. (1994). "Bcl-2 is upregulated at the CD4+ CD8+ stage during positive selection and promotes thymocyte differentiation at several control points." Immunity 1(3): 197-205.

Liston, A., K. Hardy, et al. (2007). "Impairment of organ-specific T cell negative selection by diabetes susceptibility genes: genomic analysis by mRNA profiling." Genome Biol 8(1): R12.

Liu, Z. G., S. W. Smith, et al. (1994). "Apoptotic signals delivered through the T-cell receptor of a T-cell hybrid require the immediate-early gene nur77." Nature 367(6460): 281-284.

Livak, F., M. Tourigny, et al. (1999). "Characterization of TCR gene rearrangements during adult murine T cell development." J Immunol 162(5): 2575-2580.

Lucattelli, M., S. Fineschi, et al. (2006). "Neurokinin-1 receptor blockade and murine lung tumorigenesis." Am J Respir Crit Care Med 174(6): 674-683.

Ma, A., J. C. Pena, et al. (1995). "Bclx regulates the survival of double-positive thymocytes." Proc Natl Acad Sci U S A 92(11): 4763-4767.

MacDonald, H. R., F. Radtke, et al. (2001). "T cell fate specification and alphabeta/gammadelta lineage commitment." Curr Opin Immunol 13(2): 219-224.

Magri, M., A. Yatim, et al. (2009). "Notch ligands potentiate IL-7-driven proliferation and survival of human thymocyte precursors." Eur J Immunol 39(5): 1231-1240.

Mandal, M., C. Borowski, et al. (2005). "The BCL2A1 gene as a pre-T cell receptor-induced regulator of thymocyte survival." J Exp Med 201(4): 603-614.

Mandal, M., K. M. Crusio, et al. (2008). "Regulation of lymphocyte progenitor survival by the proapoptotic activities of Bim and Bid." Proc Natl Acad Sci U S A 105(52): 20840-20845.

Mao, C., E. G. Tili, et al. (2007). "Unequal contribution of Akt isoforms in the double-negative to double-positive thymocyte transition." J Immunol 178(9): 5443-5453.

Maraskovsky, E., L. A. O'Reilly, et al. (1997). "Bcl-2 can rescue T lymphocyte development in interleukin-7 receptor-deficient mice but not in mutant rag-1-/- mice." Cell 89(7): 1011-1019.

Marino, J. H., C. Tan, et al. (2010). "Differential IL-7 responses in developing human thymocytes." Hum Immunol 71(4): 329-333.

Marsden, V. S., L. O'Connor, et al. (2002). "Apoptosis initiated by Bcl-2-regulated caspase activation independently of the cytochrome c/Apaf-1/caspase-9 apoptosome." Nature 419(6907): 634-637.

Martin, S. and M. J. Bevan (1997). "Antigen-specific and nonspecific deletion of immature cortical thymocytes caused by antigen injection." Eur J Immunol 27(10): 2726-2736.

McCarty, N., S. Paust, et al. (2005). "Signaling by the kinase MINK is essential in the negative selection of autoreactive thymocytes." Nat Immunol 6(1): 65-72.

McCaughtry, T. M., T. A. Baldwin, et al. (2008). "Clonal deletion of thymocytes can occur in the cortex with no involvement of the medulla." J Exp Med 205(11): 2575-2584.

Michie, A. M. and J. C. Zuniga-Pflucker (2002). "Regulation of thymocyte differentiation: pre-TCR signals and beta-selection." Semin Immunol 14(5): 311-323.

Mingueneau, M., W. Jiang, et al. (2012). "Thymic negative selection is functional in NOD mice." J Exp Med 209(3): 623-637.

Mody, N., J. Leitch, et al. (2001). "Effects of MAP kinase cascade inhibitors on the MKK5/ERK5 pathway." FEBS Lett 502(1-2): 21-24.

Mombaerts, P., A. R. Clarke, et al. (1992). "Mutations in T-cell antigen receptor genes alpha and beta block thymocyte development at different stages." Nature 360(6401): 225-231.

Murakami, K., N. Liadis, et al. (2010). "Caspase 3 is not essential for the induction of anergy or multiple pathways of CD8+ T-cell death." Eur J Immunol 40(12): 3372-3377.

Murphy, L. O., S. Smith, et al. (2002). "Molecular interpretation of ERK signal duration by immediate early gene products." Nat Cell Biol 4(8): 556-564.

Newton, K., A. W. Harris, et al. (1998). "A dominant interfering mutant of FADD/MORT1 enhances deletion of autoreactive thymocytes and inhibits proliferation of mature T lymphocytes." EMBO J 17(3): 706-718.

Newton, K., A. W. Harris, et al. (2000). "FADD/MORT1 regulates the pre-TCR checkpoint and can function as a tumour suppressor." EMBO J 19(5): 931-941.

Ogawara, Y., S. Kishishita, et al. (2002). "Akt enhances Mdm2-mediated ubiquitination and degradation of p53." J Biol Chem 277(24): 21843-21850.

Opferman, J. T., A. Letai, et al. (2003). "Development and maintenance of B and T lymphocytes requires antiapoptotic MCL-1." Nature 426(6967): 671-676.

Pages, G., S. Guerin, et al. (1999). "Defective thymocyte maturation in p44 MAP kinase (Erk 1) knockout mice." Science 286(5443): 1374-1377.

Pallard, C., A. P. Stegmann, et al. (1999). "Distinct roles of the phosphatidylinositol 3-kinase and STAT5 pathways in IL-7-mediated development of human thymocyte precursors." Immunity 10(5): 525-535.

Pancer, Z. and M. D. Cooper (2006). "The evolution of adaptive immunity." Annu Rev Immunol 24: 497-518.

Pellegrini, M., P. Bouillet, et al. (2004). "Loss of Bim increases T cell production and function in interleukin 7 receptor-deficient mice." J Exp Med 200(9): 1189-1195.

Peschon, J. J., P. J. Morrissey, et al. (1994). "Early lymphocyte expansion is severely impaired in interleukin 7 receptor-deficient mice." J Exp Med 180(5): 1955-1960.

Purton, J. F., Y. Zhan, et al. (2002). "Glucocorticoid receptor deficient thymic and peripheral T cells develop normally in adult mice." Eur J Immunol 32(12): 3546-3555.

Putcha, G. V., S. Le, et al. (2003). "JNK-mediated BIM phosphorylation potentiates BAX-dependent apoptosis." Neuron 38(6): 899-914.

Rajpal, A., Y. A. Cho, et al. (2003). "Transcriptional activation of known and novel apoptotic pathways by Nur77 orphan steroid receptor." EMBO J 22(24): 6526-6536.

Rathmell, J. C., T. Lindsten, et al. (2002). "Deficiency in Bak and Bax perturbs thymic selection and lymphoid homeostasis." Nat Immunol 3(10): 932-939.

Ribeiro, G. F., M. Corte-Real, et al. (2006). "Characterization of DNA damage in yeast apoptosis induced by hydrogen peroxide, acetic acid, and hyperosmotic shock." Mol Biol Cell 17(10): 4584-4591.

Rincon, M., A. Whitmarsh, et al. (1998). "The JNK pathway regulates the In vivo deletion of immature CD4(+)CD8(+) thymocytes." J Exp Med 188(10): 1817-1830.

Rodriguez-Borlado, L., D. F. Barber, et al. (2003). "Phosphatidylinositol 3-kinase regulates the CD4/CD8 T cell differentiation ratio." J Immunol 170(9): 4475-4482.

Sabapathy, K., Y. Hu, et al. (1999). "JNK2 is required for efficient T-cell activation and apoptosis but not for normal lymphocyte development." Curr Biol 9(3): 116-125.

Sabapathy, K., T. Kallunki, et al. (2001). "c-Jun NH2-terminal kinase (JNK)1 and JNK2 have similar and stage-dependent roles in regulating T cell apoptosis and proliferation." J Exp Med 193(3): 317-328.

Saint-Ruf, C., M. Panigada, et al. (2000). "Different initiation of pre-TCR and gammadeltaTCR signalling." Nature 406(6795): 524-527.

Saint-Ruf, C., K. Ungewiss, et al. (1994). "Analysis and expression of a cloned pre-T cell receptor gene." Science 266(5188): 1208-1212.

Sasaki, T., J. Irie-Sasaki, et al. (2000). "Function of PI3Kgamma in thymocyte development, T cell activation, and neutrophil migration." Science 287(5455): 1040-1046.

Schmitt, T. M. and J. C. Zuniga-Pflucker (2002). "Induction of T cell development from hematopoietic progenitor cells by delta-like-1 in vitro." Immunity 17(6): 749-756.

Schmitz, I., L. K. Clayton, et al. (2003). "Gene expression analysis of thymocyte selection in vivo." Int Immunol 15(10): 1237-1248.

Sentman, C. L., J. R. Shutter, et al. (1991). "bcl-2 inhibits multiple forms of apoptosis but not negative selection in thymocytes." Cell 67(5): 879-888.

Sidman, C. L., J. D. Marshall, et al. (1992). "Transgenic T cell receptor interactions in the lymphoproliferative and autoimmune syndromes of lpr and gld mutant mice." Eur J Immunol 22(2): 499-504.

Simon, A. K., O. Williams, et al. (2001). "Tumor necrosis factor-related apoptosis-inducing ligand in T cell development: sensitivity of human thymocytes." Proc Natl Acad Sci U S A 98(9): 5158-5163.

Singer, G. G. and A. K. Abbas (1994). "The fas antigen is involved in peripheral but not thymic deletion of T lymphocytes in T cell receptor transgenic mice." Immunity 1(5): 365-371.

Smith, K. G., A. Strasser, et al. (1996). "CrmA expression in T lymphocytes of transgenic mice inhibits CD95 (Fas/APO-1)-transduced apoptosis, but does not cause lymphadenopathy or autoimmune disease." EMBO J 15(19): 5167-5176.

Sohn, S. J., G. M. Lewis, et al. (2008). "Non-redundant function of the MEK5-ERK5 pathway in thymocyte apoptosis." EMBO J 27(13): 1896-1906.

Song, G., G. Ouyang, et al. (2005). "The activation of Akt/PKB signaling pathway and cell survival." J Cell Mol Med 9(1): 59-71.

Starr, T. K., S. C. Jameson, et al. (2003). "Positive and negative selection of T cells." Annu Rev Immunol 21: 139-176.

Stasik, I., A. Rapak, et al. (2007). "Ionomycin-induced apoptosis of thymocytes is independent of Nur77 NBRE or NurRE binding, but is accompanied by Nur77 mitochondrial targeting." Biochim Biophys Acta 1773(9): 1483-1490.

Strasser, A., A. W. Harris, et al. (1991). "bcl-2 transgene inhibits T cell death and perturbs thymic self-censorship." Cell 67(5): 889-899.

Suen, A. Y. and T. A. Baldwin (2012). "Proapoptotic protein Bim is differentially required during thymic clonal deletion to ubiquitous versus tissue-restricted antigens." Proc Natl Acad Sci U S A 109(3): 893-898.

Sugawara, T., T. Moriguchi, et al. (1998). "Differential roles of ERK and p38 MAP kinase pathways in positive and negative selection of T lymphocytes." Immunity 9(4): 565-574.

Sun, Z., C. W. Arendt, et al. (2000). "PKC-theta is required for TCR-induced NF-kappaB activation in mature but not immature T lymphocytes." Nature 404(6776): 402-407.

Sun, Z., D. Unutmaz, et al. (2000). "Requirement for RORgamma in thymocyte survival and lymphoid organ development." Science 288(5475): 2369-2373.

Swainson, L., S. Kinet, et al. (2005). "Glucose transporter 1 expression identifies a population of cycling CD4+ CD8+ human thymocytes with high CXCR4-induced chemotaxis." Proc Natl Acad Sci U S A 102(36): 12867-12872.

Swat, W., V. Montgrain, et al. (2006). "Essential role of PI3Kdelta and PI3Kgamma in thymocyte survival." Blood 107(6): 2415-2422.

Sytwu, H. K., R. S. Liblau, et al. (1996). "The roles of Fas/APO-1 (CD95) and TNF in antigen-induced programmed cell death in T cell receptor transgenic mice." Immunity 5(1): 17-30.

Takahama, Y., E. W. Shores, et al. (1992). "Negative selection of precursor thymocytes before their differentiation into CD4+CD8+ cells." Science 258(5082): 653-656.

Teague, T. K., C. Tan, et al. (2010). "CD28 expression redefines thymocyte development during the pre-T to DP transition." Int Immunol 22(5): 387-397.

Thompson, J., M. L. Burger, et al. (2010). "Protein kinase C regulates mitochondrial targeting of Nur77 and its family member Nor-1 in thymocytes undergoing apoptosis." Eur J Immunol 40(7): 2041-2049.

Thompson, J. and A. Winoto (2008). "During negative selection, Nur77 family proteins translocate to mitochondria where they associate with Bcl-2 and expose its proapoptotic BH3 domain." J Exp Med 205(5): 1029-1036.

Trampont, P. C., A. C. Tosello-Trampont, et al. (2010). "CXCR4 acts as a costimulator during thymic beta-selection." Nat Immunol 11(2): 162-170.

Tsujimoto, Y. and S. Shimizu (2000). "VDAC regulation by the Bcl-2 family of proteins." Cell Death Differ 7(12): 1174-1181.

Vacchio, M. S., V. Papadopoulos, et al. (1994). "Steroid production in the thymus: implications for thymocyte selection." J Exp Med 179(6): 1835-1846.

Van De Wiele, C. J., J. H. Marino, et al. (2004). "Thymocytes between the beta-selection and positive selection checkpoints are nonresponsive to IL-7 as assessed by STAT-5 phosphorylation." J Immunol 172(7): 4235-4244.

Van de Wiele, C. J., J. H. Marino, et al. (2007). "Impaired thymopoiesis in interleukin-7 receptor transgenic mice is not corrected by Bcl-2." Cell Immunol 250(1-2): 31-39.

Veis, D. J., C. M. Sorenson, et al. (1993). "Bcl-2-deficient mice demonstrate fulminant lymphoid apoptosis, polycystic kidneys, and hypopigmented hair." Cell 75(2): 229-240.

von Freeden-Jeffry, U., N. Solvason, et al. (1997). "The earliest T lineage-committed cells depend on IL-7 for Bcl-2 expression and normal cell cycle progression." Immunity 7(1): 147-154.

Walsh, C. M., B. G. Wen, et al. (1998). "A role for FADD in T cell activation and development." Immunity 8(4): 439-449.

Wang, A., J. Rud, et al. (2009). "Phosphorylation of Nur77 by the MEK-ERK-RSK cascade induces mitochondrial translocation and apoptosis in T cells." J Immunol 183(5): 3268-3277.

Wang, X., Y. Zhang, et al. (2009). "c-Fos enhances the survival of thymocytes during positive selection by upregulating Bcl-2." Cell Res 19(3): 340-347.

Wange, R. L. (2000). "LAT, the linker for activation of T cells: a bridge between T cell-specific and general signaling pathways." Sci STKE 2000(63): re1.

Werlen, G., B. Hausmann, et al. (2000). "A motif in the alphabeta T-cell receptor controls positive selection by modulating ERK activity." Nature 406(6794): 422-426.

Whitfield, J., S. J. Neame, et al. (2001). "Dominant-negative c-Jun promotes neuronal survival by reducing BIM expression and inhibiting mitochondrial cytochrome c release." Neuron 29(3): 629-643.

Williams, O., T. Norton, et al. (1998). "The action of Bax and bcl-2 on T cell selection." J Exp Med 188(6): 1125-1133.

Wilson, A., H. R. MacDonald, et al. (2001). "Notch 1-deficient common lymphoid precursors adopt a B cell fate in the thymus." J Exp Med 194(7): 1003-1012.

Wojciechowski, S., P. Tripathi, et al. (2007). "Bim/Bcl-2 balance is critical for maintaining naive and memory T cell homeostasis." J Exp Med 204(7): 1665-1675.

Wolfer, A., A. Wilson, et al. (2002). "Inactivation of Notch1 impairs VDJbeta rearrangement and allows pre-TCR-independent survival of early alpha beta Lineage Thymocytes." Immunity 16(6): 869-879.

Woronicz, J. D., B. Calnan, et al. (1994). "Requirement for the orphan steroid receptor Nur77 in apoptosis of T-cell hybridomas." Nature 367(6460): 277-281.

Xi, H. and G. J. Kersh (2004). "Sustained early growth response gene 3 expression inhibits the survival of CD4/CD8 double-positive thymocytes." J Immunol 173(1): 340-348.

Xi, H., R. Schwartz, et al. (2006). "Interplay between RORgammat, Egr3, and E proteins controls proliferation in response to pre-TCR signals." Immunity 24(6): 813-826.

Xia, Z., M. Dickens, et al. (1995). "Opposing effects of ERK and JNK-p38 MAP kinases on apoptosis." Science 270(5240): 1326-1331.

Yashiro-Ohtani, Y., Y. He, et al. (2009). "Pre-TCR signaling inactivates Notch1 transcription by antagonizing E2A." Genes Dev 23(14): 1665-1676.

Yu, Q., B. Erman, et al. (2004). "IL-7 receptor signals inhibit expression of transcription factors TCF-1, LEF-1, and RORgammat: impact on thymocyte development." J Exp Med 200(6): 797-803.

Yuan, J., R. B. Crittenden, et al. (2010). "c-Myb promotes the survival of CD4+CD8+ double-positive thymocytes through upregulation of Bcl-xL." J Immunol 184(6): 2793-2804.

Zhang, J., K. Mikecz, et al. (2000). "Spontaneous thymocyte apoptosis is regulated by a mitochondrion-mediated signaling pathway." J Immunol 165(6): 2970-2974.

Zhang, N. and Y. W. He (2005). "The antiapoptotic protein Bcl-xL is dispensable for the development of effector and memory T lymphocytes." J Immunol 174(11): 6967-6973.

Zhou, T., J. Cheng, et al. (1996). "Inhibition of Nur77/Nurr1 leads to inefficient clonal deletion of self-reactive T cells." J Exp Med 183(4): 1879-1892.

Zucchelli, S., P. Holler, et al. (2005). "Defective central tolerance induction in NOD mice: genomics and genetics." Immunity 22(3): 385-396.

Chapter 8

Drug Resistance and Molecular Cancer Therapy: Apoptosis Versus Autophagy

Rebecca T. Marquez, Bryan W. Tsao, Nicholas F. Faust and Liang Xu

Additional information is available at the end of the chapter

1. Introduction

The majority of chemo/radiotherapies inhibit cancer cell growth by activating cell death pathways, such as apoptosis, necrosis, and autophagy-associated cell death. However, as the disease progresses, cancer cells can acquire a variety of genetic and epigenetic alterations, which leads to dysregulation of cell death-associated signaling pathways and chemo/radioresistance. Designing novel drugs and enhancing therapeutic strategies to improve survival and quality of life for cancer patients must specifically target pathways responsible for drug resistance. Two cellular mechanisms can contribute to chemo/radioresistance: inhibition of apoptotic cell death pathways and induction of autophagy, a cell survival response. The development of novel drugs and extensive research studies has provided significant insight into the aberrant regulation of apoptosis and key apoptosis inhibitor proteins during tumorigenesis. However, the extensive dysregulation of cell growth pathways in cancer cells makes it necessary to target multiple pathways in order to elicit a lasting death response. Autophagy, classically designated as a cell "survival" mechanism, appears to play a greater role in cell death than previously conceived. This contradiction between autophagy-associated cell survival versus cell death has intensified the interest in this field of research in cancer therapeutics. Understanding how autophagic cells cross the threshold from cell survival to cell death during drug treatments is imperative for identifying more potent therapies. Utilizing novel treatments that will re-activate apoptotic cell death pathways, while driving autophagy-associated cell death will lead to more effective chemotherapies, thereby enhancing overall patient survival.

Keywords: apoptosis; autophagy; programmed cell death; molecular therapy; personalized medicine; signaling pathways

2. Apoptosis pathways

Cancer cells can acquire apoptosis-resistance during treatment by up-regulating multiple pro-survival factors, such as inhibitors of apoptosis proteins (IAPs), nuclear factor-κB (NF-κB), and the B cell CLL/lymphoma-2 (BCL-2) family proteins. There are two major apoptosis signaling pathways, the extrinsic and intrinsic apoptosis signaling pathways (Figure 1). The extrinsic (death receptor) apoptosis pathway is induced by the binding of cell death ligands, such as FAS ligand or TNF to cell death receptors, FAS receptor or tumor necrosis factor receptor,TNFR1, respectively. Activation of these death receptors results in caspase 8 activated cell death [1]. The intrinsic (mitochondrial or BCL-2 regulated) apoptosis pathway can be activated by cellular stresses or chemo/radiotherapies that lead to functional activation of the pro-apoptotic BCL-2 family proteins, which induce mitochondrial outer membrane permeabilitization (MOMP) and cytochrome *c* release into the cytosol. Once in the cytosol, cytochrome *c* induces formation of the apoptosome complex, which contains cytochrome *c*, caspase 9 and apoptotic protease-activating factor-1 (APAF-1), followed by activation of downstream caspases 3, and 7[2]. While the intrinsic apoptosis pathway is considered to be regulated by BCL-2, the extrinsic pathway can also be regulated by BCL-2 family members via crosstalk with the intrinsic pathway. This crosstalk occurs through caspase 8 cleavage and activation of the BH3-interacting domain death agonist (BID). The cleavage product, truncated BID (tBID), is required for death receptor-induced apoptosis in some cell types. During tumorigenesis, both the extrinsic and intrinsic apoptosis signaling pathways become dramatically dysregulated thereby leading to increased cell survival upon chemo/radiotherapy. This chapter will discuss exploitation of factors regulating apoptosis, such as second mitochondria-derived activator of caspases (SMAC) and BH3-only proteins, as molecular targets utilized to overcome apoptosis resistance in cancer cells.

3. IAP family proteins promote apoptosis-resistance

IAPs are a pivotal class of pro-survival factors that suppress apoptosis against a large variety of apoptotic stimuli, including chemotherapeutic agents, radiation, and immunotherapy in cancer cells[3-5]. Elevated expression of IAPs is a common occurrence in multiple cancer types, while eliciting a wide range of biological responses that promote cancer cell survival and proliferation[6]. Therefore, IAPs are attractive molecular targets for anti-cancer therapies in order to decrease apoptosis-resistance, thereby enhancing cancer therapeutics and increasing patient survival.

IAPs are characterized by baculoviral IAP repeat (BIR) domains, which are required for the majority of IAP-mediated protein-protein interactions and inhibition of apoptosis[7]. Eight IAPs have currently been identified in humans, but the most studied IAP members include the X chromosome-linked IAP protein (XIAP), cellular IAP1 (cIAP-1), and cellular IAP2 (cIAP-2)[8]. IAPs inhibit both the intrinsic and extrinsic apoptotic pathways (Figure 1). XIAP binds to and inhibit caspases 3, 7, and 9, while cIAPs negatively regulate caspase 8 activation through TNFR1 signaling[9]. IAPs also contain a carboxyl-terminal RING domain, which enables them to function as E3 ubiquitin ligases[6]. XIAP and cIAPs can promote cancer cell survival and proliferation by inhibiting caspase activation, IAP-antagonist binding, or by acting as critical mediators of the NF-κB pathway.

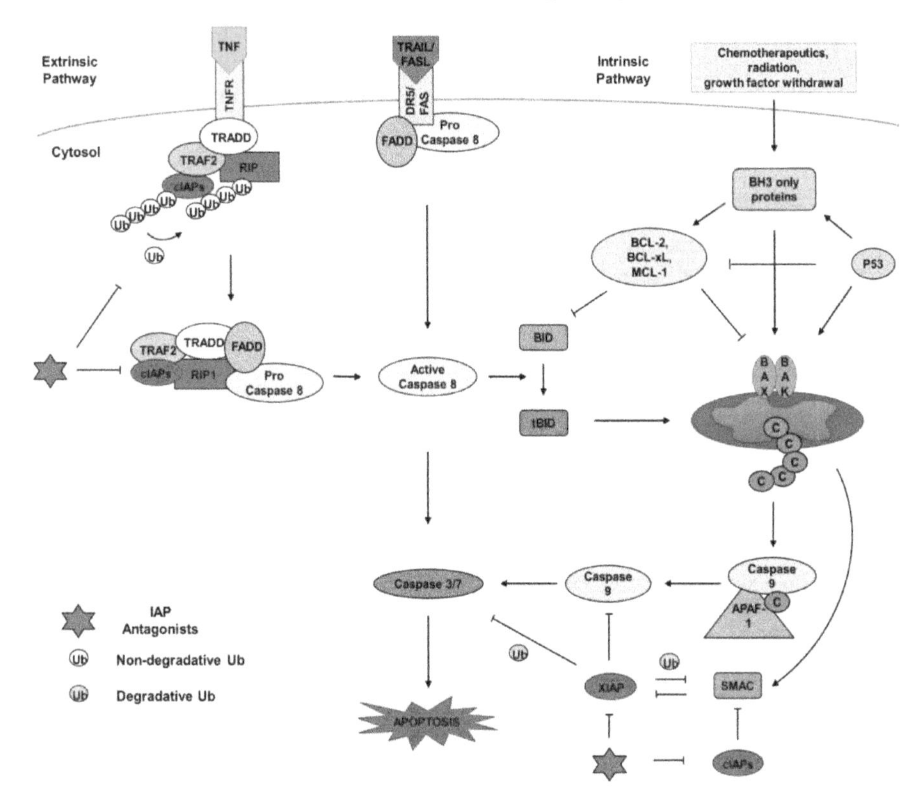

Figure 1. Extrinsic and intrinsic apoptosis signaling pathways. The extrinsic (death receptor) apoptosis pathway is induced by the binding of cell death ligands, TNF, FASL or TRAIL, to cell death receptors TNFR, FAS, or DR5, respectively. Activation of the death receptors results in caspase 8 activated cell death. The intrinsic (mitochondrial or BCL-2 regulated) apoptosis pathway can be activated by cellular stresses or chemo/radiotherapies. This leads to functional activation of the pro-apoptotic BCL-2 family proteins which induces cytochrome *c* or SMAC release into the cytosol. Cytochrome *c* induces formation of the apoptosome complex, which contains cytochrome *c*, caspase 9, and APAF-1, followed by activation of downstream caspase 3 and 7. SMAC can promote apoptosis by binding to XIAP, which results in the subsequent release of caspase 9 and downstream activation of apoptosis. cIAPs are capable of inhibiting SMAC by blocking this interaction. The crosstalk between the extrinsic and intrinsic pathways occurs through caspase 8 cleavage and activation of the BID. The cleavage product, tBID, is required for death receptor-induced apoptosis in some cell types. During tumorigenesis, both the extrinsic and intrinsic apoptosis signaling pathways become dramatically dysregulated thereby leading to increased cell survival during chemo/radiotherapy. IAP antagonists can inhibit the anti-apoptotic actions of XIAP and cIAPs in both the intrinsic and extrinsic apoptosis pathways.

4. XIAP is a potent caspase inhibitor

XIAP protein is the first well-characterized IAP family member[4, 10, 11]. XIAP is overexpressed in approximately 25% of the 60 NCI human cancer cell lines and can predict response to chemotherapy[12-16]. Although it was initially believed that all IAP proteins blocked apoptosis by directly binding caspases, it was later found that only XIAP directly binds to and inhibits caspases 3, 7, and 9 (Figure 1)[10, 17, 18]. As caspase 9 is an initiator caspase, it is considered the most critical target for XIAP's anti-apoptotic function[19]. Structural studies have outlined the protein interactions utilized by XIAP to inhibit caspase function. The BIR3 domain of XIAP binds to the catalytic domain of caspase 9 while the linker region between XIAP BIR1 and BIR2 binds to caspase 3 or 7[20, 21],[22]. In addition to binding and blocking caspase catalytic sites, XIAP also utilizes its E3 ubiquitin ligase function for targeting and ubiquitylating caspase 3 for proteasome degradation[23]. Therefore, due to XIAP's ability to inhibit multiple caspases, either directly or via ubiquitylation, XIAP has become a premiere molecular target for current chemotherapies (Figure 1).

5. cIAP regulation of the NF-κB signaling pathway

Although IAPs are typically known to bind and inhibit caspases, cIAPs also modulate ubiquitin-dependent signaling events of the extrinsic apoptosis pathway and regulate activation of NF-κB[24]. cIAPs are required for stimulus-dependent activation of the canonical pathway and for constitutive suppression of the non-canonical NF-κB pathway (Figure 2)[8]. NF-κB is a transcription factor involved in angiogenesis, metastasis, and cell proliferation[8]. Upon activation, NF-κB regulates transcription of pro-survival genes such as TNFα, cIAPs, BCL-2 and other apoptosis-related proteins. Furthermore, blocking NF-κB pathway can sensitize cancer cells to chemotherapeutic agents and radiation[25-27].

In the canonical NF-κB pathway, the inhibitor of NF-κB (IκBα) binds to NF-κB, thereby preventing NF-κB nuclear translocation from the cytoplasm into the nucleus in unstimulated cells[28]. TNF-mediated activation of NF-κB requires the assembly of an ubiquitin-dependent signaling complex[29]. TNF ligand binding to TNFR1 induces the formation of a signaling complex by initially recruiting TNFR1-associated death domain protein (TRADD) and TNFR-associated factor 2 (TRAF2), followed by recruitment of receptor-interacting protein 1 (RIP1) and c-IAP proteins (Figure 2)[30, 31]. Within this complex, cIAPs promote nondegradative polyubiquitylation of RIP1, in addition to themselves, to generate a binding platform for assembly of the IκB kinase (IKK) complex[32-34]. This leads to the activation of IKKβ, which results in phosphorylation of IκBα, prompting IκBα polyubiquitylation and subsequent degradation. This allows NF-κB to translocate to the nucleus and activate target genes[28]. Therefore, cIAPs positively regulate the canonical NF-κB pathway.

Alternatively, in the non-canonical pathway, cIAPs negatively regulate NF-κB transcription by ubiquitylating and targeting NF-κB-inducing kinase (NIK) for proteasomal degradation[35]. In unstimulated cells, a cytoplasmic complex composed of cIAPs, TRAF2, TRAF3 and NIK, maintains constitutive ubiquitin-dependent proteasomal degradation of

NIK (Figure 2) [35-41]. Accumulation of NIK is acquired by dissociation of this cytoplasmic complex. Upon ligand binding, receptors of the TNFR family, such as CD40, recruit TRAF2, TRAF3 and the cIAP proteins into their respective signaling complexes. This results in cIAP ubiquitylation and degradation of the cIAPs, TRAF2, and TRAF3, which leads to stabilization and accumulation of NIK and downstream activation of NF-κB anti-apoptotic target genes[9, 42]. The conflicting roles that cIAPs play in inducing or inhibiting NF-κB signaling pathway display an additional layer of complexity when developing therapeutic drugs targeting cIAPS.

6. SMAC: IAP-antagonist

SMAC is a regulator of the intrinsic apoptosis pathway and becomes released from the mitochondria upon mitochondrial outer membrane permeabilitization (MOMP) (Figure 1). Structural studies show that SMAC induces apoptosis by binding to and sequestering IAPs from binding to caspases[43-45]. As previously mentioned, the BIR3 domain of XIAP binds to the N-terminus of small subunit p12 of processed caspase 9. SMAC protein contains a region homologous to the caspase 9 p12 subunit, therefore, it can also bind to XIAP BIR3 domain[20]. SMAC binding of XIAP allows the subsequent release of caspase 9 and activation of downstream signaling leading to apoptosis[46]. While cIAPs are not potent inhibitors of caspases, cIAPs are able to bind to SMAC with high affinity, thereby preventing SMAC from disrupting XIAP-mediated inhibition of caspases[6].

6.1. The role of IAP and SMAC and clinical outcome

Due to the importance of apoptosis resistance during chemo/radiotherapy, the expression of IAP proteins and IAP inhibiting proteins, such as SMAC, have demonstrated significant correlation with clinicopathological data[6, 47]. Altered expression of cIAPs in cancer cells is typically due to chromosomal aberrations, such as genomic ampifications, translocations and deletions. Genomic amplification at the 11q21-q23 genomic loci of both cIAP1 and cIAP2 has been detected in many cancers, including esophageal squamous cell carcinomas, liver cancer, lung cancer, and cervical cancer[48-51]. Furthermore, immunohistochemical analysis of cervical cancers from patients treated only with radiotherapy had high levels of nuclear cIAP1 staining and demonstrated that both overall survival and local recurrence-free survival was significantly poorer compared to patients with little or no nuclear cIAP1[50]. Genomic translocations, such as t(11;18)(q21;q21), results in the fusion of the BIR domains of cIAP2 with paracaspase mucosa-associated lymphoid tissue lymphoma translocation protein 1(MALT1) and occurs frequently in mucosa-associated lymphoid tissues[52-54]. The resulting cIAP2-MALT1 fusion protein constitutively activates the NF-κB signaling pathway[53, 55].

As previously discussed, cIAPs act as oncogenes in most cancers, however, cIAPs in multiple myeloma has demonstrated tumor suppressive properties. In multiple myeloma, chromosomal deletions of cIAP-1/2 resulted in stabilization of NIK, which induced constitutive aberrant activation of the non-canonical survival NF-κB pathway[37, 39]. This further delineates the important balancing act of cIAPs in regulating the NF- κB pathways and cell survival.

XIAP expression is also dysregulated in many cancers and correlates with clinical outcome[6]. XIAP is upregulated in clear-cell renal cell carcinoma and correlates with increasing tumor stage, dedifferentiation, and aggressive growth[56]. XIAP was also shown to be an independent prognostic marker for non-muscular invasive bladder cancer, colon cancer, and liver cancer[57-59]. In invasive breast ductal carcinoma, nuclear expression of XIAP correlated with shortened overall survival[60]. Interestingly, a prostate cancer study showed patients with high XIAP levels had a much lower probability of tumor recurrence than those with lower XIAP expression. Furthermore, patients with high-grade prostate tumors who had high XIAP levels had a lower risk of recurrence compared with patients whose tumors express low XIAP[61]. This demonstrates that while many cancers have a correlation with high XIAP expression levels and poor prognosis, some cancers have additional altered mechanisms associated with poor clinical outcome. This further supports the need for tumor expression profiling in order to determine whether an individual's tumor is apoptosis-resistant. Pre-treatment screening will allow physicians to identify the proper treatment regimen in order to avoid unnecessary toxicity and relapse.

The down-regulation of IAP inhibitor, SMAC, has also been shown to play a significant role in inhibiting IAPs in cancer and correlates with clinical outcome[6]. In rectal cancer, high expression levels of SMAC correlated with 5-year recurrence free survival rate and 5-year local relapse-free survival rate[62]. Down-regulation of SMAC has been shown to be associated with disease progression in many cancer types, such as lung, hepatocellular carcinoma, testicular cancer[63-65]. In renal cell carcinoma, low levels of SMAC correlated with advanced tumor stage, poor prognosis, and a reduced probability of recurrence-free survival[56, 66]. Furthermore, XIAP expression increased with stage and grade, while mRNA and protein expression levels of SMAC did not significantly change. This results in a relative increase of anti-apoptotic XIAP over pro-apoptotic SMAC, thereby contributing to apoptosis resistance in renal cell carcinoma[66].

6.2. IAP antagonists as therapy to overcome apoptosis-resistance

Due to the dysregulation and contribution of IAPs towards chemo/radioresistance, researchers have developed several targeting strategies, such as small-molecule IAP antagonists, including SMAC mimetics, and antisense oligonucleotides. Table 1 shows a subset of IAP antagonists currently used in clinical trials.

6.3. SMAC-mimetics

Several studies have shown that overexpression of SMAC sensitizes neoplastic cells to apoptotic cell death[67, 68]. Therefore, SMAC mimetics have been developed in order to sensitize cancer cells to apoptotic stimuli, such as chemo/radiotherapy. Synthetic SMAC N-terminal peptides fused to cell-permeabilizing peptides were initially used as SMAC mimetics for treating cancer cells. These peptides were found to bypass mitochondrial regulation and sensitize both human cancer cells in culture and tumor xenographs in mice to apoptosis when combined with TNF-related apoptosis-inducing ligand (TRAIL) or chemotherapeutic drug treatments[69, 70]. While appearing effective, SMAC peptides did not possess good pharmacological properties and, therefore, could not be used as therapeutic agents. Researchers then utilized 3D structure analysis of SMAC bound to XIAP BIR3 domain to design and synthesize small molecule SMAC mimetics[71-73]. These compounds show at least 20-fold enhanced binding to XIAP BIR3 domain over the natural SMAC peptide in a cell-free system[72-74]. Small molecule SMAC mimetics also bind and inhibit cIAP-1 and cIAP-2 activities and promote apoptosis synergistically with proapoptotic stimuli, such as TRAIL or TNFα, in cancer cells that were previously determined to be resistant to TRAIL or TNFα[71].

Drug	**Cancer type(s)**	**Clinical Trial**	**Co-therapy**	**Outcome**
AT-406	Solid tumors, lymphoma	Phase 1	None	Ongoing.[225]
	AML	Phase 1	Daunorubicin and Cytarabine	Ongoing.[225]
AEG35156	AML	Phase 1/2	High-dose Cytarabine and Idarubicin	AEG35156 treatment led to dose-dependent decreases of XIAP mRNA and protein levels. Apoptosis induction was detected.[195]
	AML	Phase 1/2	Cytarabine and Idarubicin	Very effective when combined with chemotherapy in patients with AML refractory to a single induction regimen.[87]
YM155	Advanced refractory solid tumors	Phase 1	None	The safety profile, plasma concentrations achieved, and antitumor activity.[209]
	NSCLC	Phase 2	None	Modest single-agent activity in patients with refractory, advanced NSCLC. A favorable safety/tolerability profile was reported.[210]

AML– Acute myeloid leukemia; Non-small cell lung cancer – NSCLC

Table 1. Selective list of IAP antagonists undergoing clinical trials with and without combination therapy.

Pre-clinical and clinical data has demonstrated that SMAC mimetics may show more therapeutic promise in combination with conventional chemotherapeutic drugs, death receptor agonist or radiation therapy (Table 1). Research from our lab demonstrated that the

SMAC mimetic, SH130, disrupts the binding between XIAP/cIAP and SMAC. Upon combination treatment, SH130 enhances ionizing radiation-induced apoptosis *in vitro* and induces 80% tumor regression in hormone-refractory prostate cancer models[75]. We also demonstrated that SMAC mimetic, SH122, can induce cell death via both the extrinsic and intrinsic apoptosis pathways. In combined treatment with death receptor ligand, TRAIL, SH122 induces TRAIL-mediated cell death in prostate cancer cell lines by blocking IAPs and NF-κB[76].

SMAC mimetics have proven tremendous efficacy when used in combination with treatments to induce apoptosis in apoptosis-resistant cells[77]. Interestingly, it was also shown that SMAC mimetic treatment alone could induce apoptosis in a subset of non-small-cell lung cancer cell lines[78]. It was later determined that autocrine-secreted TNFα-mediated apoptosis signals that were inhibited by IAP proteins. Treatment with the SMAC mimetic promoted formation of RIP1-dependent caspase 8-activating complex leading to apoptosis in these cells[78]. It has also been demonstrated that SMAC mimetic binding of cIAPs leads to rapid ubiquitination and proteasomal degradation of cIAPs[35]. Therefore, in addition to targeting XIAP to relieve caspase 9 inhibition in the intrinsic cell death pathway, SMAC mimetics can induce cIAPs auto-ubiquitination and degradation, which leads to NF-κB activation and TNFα secretion. The autocrine TNFα signaling in turn induces caspase 8 activation and cancer cell death (Figure 2).

6.4. cIAP- and XIAP-selective antagonists

SMAC mimetics have broad specificity by inhibiting both XIAP and cIAPs. Currently, the individual roles of IAPs in apoptosis resistance, as well as BIR domain structure, are unexplored. Therefore, more selective antagonists are designed in order to provide greater specificity for the diverse IAPs. CS3 is a cIAP1/2 selective antagonist and has been shown to induce degradation of cIAP1/2, activate canonical, non-canonical NF-κB signaling pathways, and induce cell death[79]. Although CS3 is capable of inducing cell death, cIAP-selective antagonists are significantly less potent in promoting apoptosis than pan-selective compounds[79].

Embelin, the active ingredient of traditional herbal medicine, is a potent IAP antagonist that binds to the XIAP BIR3 domain. We have shown that embelin inhibits cell growth, induces apoptosis, and activates caspase 9 in prostate cancer cells with high levels of XIAP, but has a minimal effect on normal prostate epithelial cells with low levels of XIAP[80]. Furthermore, embelin combined with radiation potently suppressed prostate cancer cell proliferation that was associated with S and G2/M cell cycle arrest[81]. Moreover, the combination treatment promoted caspase-independent apoptosis. *In vivo*, embelin significantly improved tumor response to x-ray radiation in PC-3 xenograft model. Combination therapy resulted in tumor growth delay and prolonged time to tumor progression, with minimal systemic toxicity. These findings demonstrate the potential to utilize embelin as a novel adjuvant therapeutic candidate for the treatment of hormone-refractory prostate cancer that is resistant to radiation therapy[81].

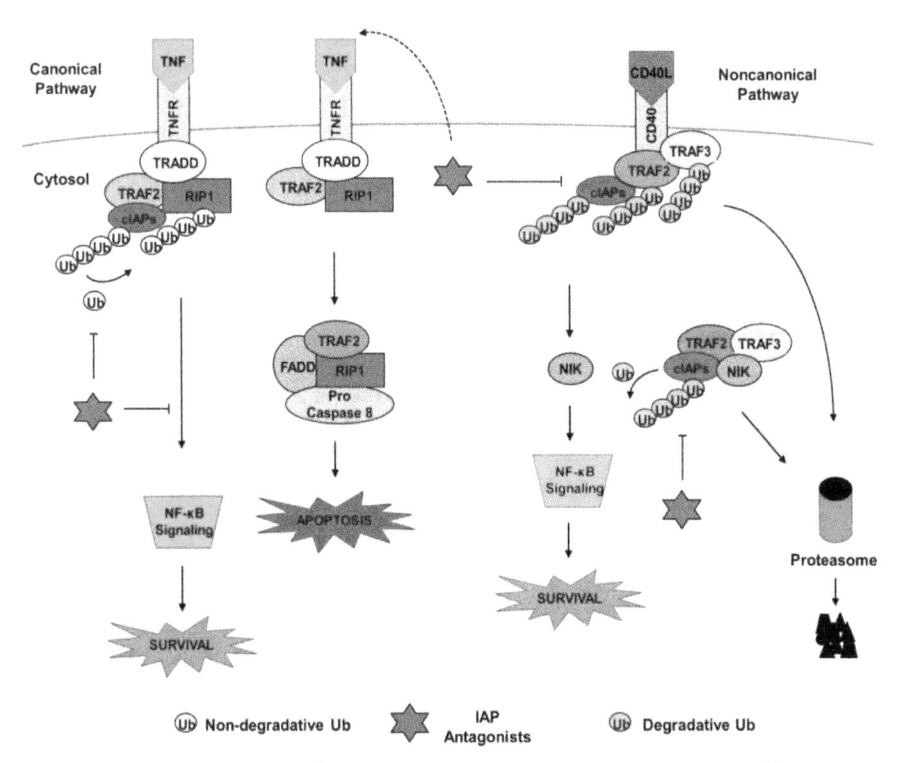

Figure 2. Canonical and non-canonical prosurvival NF-κB pathways. cIAPs are required for stimulus-dependent activation of NF-κB canonical pathway and alternatively for constitutive suppression of the non-canonical NF-κB pathway. TNF-mediated activation of the canonical NF-κB pathway requires the assembly of an ubiquitin-dependent signaling complex comprised of TRADD, TRAF2, RIP1, and cIAPs. cIAPs induce non-degradative ubiquitylation of both RIP1 as well as themselves which leads to activation of downstream pro-survival NF-κB signaling. IAP antagonists can inhibit NF-κB canonical pathway by preventing cIAP ubiquitylation of RIP1 which leads to recruitment of pro-caspase 8, thereby inducing apoptosis. Alternatively, in the non-canonical pathway, cIAPs negatively regulate NF-κB transcription by ubiquitylating and targeting NIK for proteasomal degradation. In unstimulated cells, a cytoplasmic complex composed of cIAPs; TRAF2, TRAF3 and NIK, maintains constitutive ubiquitin-dependent proteasomal degradation of NIK, thereby preventing activation of NF-κB pathway. Upon ligand binding, receptors of the TNFR family, such as CD40, recruit TRAF2, TRAF3 and the cIAP proteins into their respective signaling complexes. This results in cIAP ubiquitylation and subsequent degradation of cIAPs, TRAF2, and TRAF3. Degradation of this complex leads to stabilization and accumulation of NIK and downstream activation of NF-κB anti-apoptotic target genes. Interestingly, IAP antagonists can switch the non-canonical NF-κB signaling pathway from pro-survival to pro-apoptotic pathway (dashed arrow). IAP antagonists induce activation of this pathway by blocking cIAP inhibition, which leads to TNFα secretion. The autocrine TNFα signaling in turn induces caspase 8 activation and cancer cell death.

IAP antagonists have proven to be effective in overcoming apoptosis resistance in cancer cells. Clinical trials are currently underway to test the applicability of small-molecule IAP antagonists in single and combined anti-cancer therapies. Preliminary results suggest that IAP antagonists are well tolerated and effective in inhibiting IAPs (Table I)[6]. In addition, these molecules are providing insight into additional regulatory networks that exist in cancer cells, thereby providing new understanding of apoptosis resistance.

6.5. Inhibition of IAPs through RNA interference

Inhibition of IAPs using RNAi has further demonstrated the role of IAPs in drug resistance. Esophageal cancer cell lines transfected with XIAP siRNA demonstrated increased cell apoptosis[82]. Another study demonstrated that RNAi targeting of XIAP increased breast and pancreatic cancer cell susceptibility to functionally diverse chemotherapeutic agents, including TRAIL and taxanes and therefore increasing the effectiveness of chemotherapeutic agents[83]. Furthermore, *in vivo* studies also demonstrated that inhibition of XIAP by RNAi radiosensitized lung cancer cells by up-regulating apoptotic signaling and down-regulating cell survival[84]. We have also shown that combination treatment using RNAi silencing of IAPs and SH122 SMAC mimetic shows a greater sensitization of cells to apoptosis, than SMAC mimetic alone[76].

Clinical trials using anti-sense oligonucleotide AEG35156 is proving to be successful. The first-in-human study with AEG35156 in patients with advanced refractory cancers demonstrated that the compound was well tolerated and showed some anti-tumor activity[83]. However, AEG35156 was less effective in Phase I clinical trials with pancreatic cancer patients[85, 86]. Phase II trials treating primary refractory AML patients with both chemotherapy and AEG35156 demonstrated a 91% rate of complete remission[87]. Therefore, RNAi therapy shows significant promise in treating apoptosis-resistant cancers. While AEG35156 demonstrates promise in treating primary refractive disease, it is important to identify the patients that express high levels of IAPs in order to gain the most therapeutic benefit. Again, this demonstrates the need for molecular marker screening in order to develop personalized therapies.

7. BCL-2 family proteins regulate the intrinsic apoptotic pathway

In addition to XIAPs, the BCL-2 protein family members are essential regulators of the intrinsic apoptotic pathway, also known as the BCL-2-regulated pathway, and significant contributors to apoptosis-resistance during chemo/radiotherapies[88]. BCL-2 family members are characterized by their BCL-2 homology (BH) domain and can be categorized into three classes: the anti-apoptotic multi-domain proteins, such as BCL-2, BCL-xL, and MCL-1, are essential for cell survival, the pro-apoptotic BH3-only proteins, such as BID, BIM, BAD, and PUMA, initiate apoptosis signaling; and the pro-apoptotic multi-domain effector proteins, such as BAX and BAK, are required for MOMP and activation of caspases that leads to cell death[89, 90]. Both the anti-apoptotic and pro-apoptotic functions of BCL-2 family members are regulated through their BH domains [91, 92]. Furthermore, the BH1-

BH3 domains of anti-apoptotic proteins form a hydrophobic binding pocket that binds the α-helix of the BH3-only pro-apoptotic proteins [93, 94]. BCL-2 proteins are typically found at the outer mitochondrial membrane (OMM), however, they can also be localized to the endoplasmic reticulum (ER) and in the cytosol[95].

Death signals induced by DNA damage, growth factor deprivation, or chemotherapies induce apoptosis via the mitochondrial pathway (Figure 1) by transcriptional or post-translational activation of BH3-only proteins[96-98]. After activation, the pro-apoptotic BH3-only proteins prompt a conformational change of monomeric BAX and BAK resulting in homo-oligomerization and activation[99-101]. Activated BAX and BAK cause MOMP, followed by release of cytochrome *c* and other pro-apoptotic factors, such as SMAC, from the mitochondria. BAX and BAK are essential for the pro-apoptotic function of BH3-only proteins, therefore, loss of BAX and BAK prevents apoptotic cell death[101, 102]. Cancer cells that display overexpression of anti-apoptotic proteins and/or down-regulation of pro-apoptotic proteins, have the potential to evade chemotherapeutic cell death resulting in drug resistance.

While it is generally accepted that activation of BAX and BAK is required to induce permeabilization of the mitochondria, there are multiple models that describe the mechanisms used in the activation/inhibition of BAK/BAX. One model suggests that BH3-only activating proteins, such as Bid or Bim, directly bind to BAX/BAK to induce oligomerization and subsequent activation [65, 103-106]. Another model describes an indirect mechanism. Anti-apoptotic proteins, such as BCL-2 and BCL-xL, inhibit cell death by binding to and sequestering activating BH3-only proteins thereby preventing their activation of BAX/BAK[107-110]. The indirect mechanism involves a subset of BH3-only proteins, called sensitizers, which induce BAX/BAK oligomerization indirectly, by binding anti-apoptotic proteins, thereby displacing the activating BH3-only proteins allowing them to bind to BAX/BAK[111, 112].

Anti-apoptotic proteins, BCL-2 and BCL-xL, are also capable of heterodimerizing with BAX or BAK, thereby inhibiting BAX or BAK[113-115]. It has been shown that BCL-2 undergoes a conformational change to bind to and inhibit oligomerization of mitochondrial membrane bound Bax. However, if BAX is in excess, apoptosis resumes due to the availability of free BAX able to activate the apoptotic pathway [115].

The activation models, as described in the previous paragraphs, are simplified examples of the complex interactions required to carry out the intrinsic apoptosis pathway. Dysregulation of intrinsic apoptosis pathways, due to altered ratios of antiapoptotic members to proapoptotic members, leads to apoptotic blocks. Identifying the proteins involved in these blocks is essential for designing more effective rational therapies. Studies called "BH3 profiling" used BH3 peptides that selectively antagonize BCL-2 family members to identify apoptotic blocks in cancer cells[107, 116]. It was demonstrated that BH3-only proteins show distinct binding preferences to anti-apoptotic BCL-2 family members[107, 116]. Identifying differential BH3-only protein binding affinities for anti-apoptotic BCL-2 protein family members has led to the development of specific small molecule inhibitors of anti-apoptotic BCL-2 proteins which are designed to overcome apoptosis resistance in cancer cells and induce cell death.

7.1. Dysregulation of the BCL-2 family of proteins augments chemo/radioresistance

The dysregulation of BCL-2 family members, such as overexpression of anti-apoptotic genes or silencing of pro-apoptotic genes, is a key determinant for apoptosis-resistance during tumorigenesis and chemotherapy. BCL-2 was initially discovered to be overexpressed in human B-cell lymphomas and is located near chromosomal translocation break points frequently found in B-cell lymphomas[118]. Additional studies have demonstrated that BCL-2 protein levels in cancers are enhanced due to promoter hypomethylation, loss of inhibitory microRNA expression, and gene amplifications, signifying that up-regulation of BCL-2 expression is often found in a variety of cancers[119, 120].

Expression of anti-apoptotic BCL-2 family members has a significant effect on chemoresistance and prognosis[120, 121]. BCL-2, BCL-xL, and MCL-1 expression increases during prostate cancer progression[122]. Furthermore, BCL-2/BCL-xL expression levels correlate with resistance to a wide spectrum of chemotherapeutic agents[123, 124]. Alternatively, the pro-apoptotic BCL-2 family members can be down-regulated resulting in suppressed apoptosis. Spontaneous deletions or mutations of BAX have been observed in colorectal tumors, which results in significant reduction of apoptosis in response to anticancer agents[125, 126]. The BH3-only protein PUMA is also down-regulated in melanoma and Burkitt lymphomas[127, 128].

Dysregulation of BCL-2 family of proteins also occurs in cancer cells due to a loss of p53 tumor suppressor expression or function. p53 expression is lost in a majority of cancers. p53 can activate transcription of BAX, BID, PUMA and NOXA (Figure 1)[97, 129-132]. Cytosolic accumulation of p53 results in activation of BAX similarly to the BH3-only activating BCL-2 proteins, thereby inducing apoptosis[133]. Interestingly, p53 has also been shown to inhibit anti-apoptotic BCL-2 family members as well. DNA damage induces p53-Bcl-2 binding, thereby sequestering BCL-2 from inhibiting BAX/BAK oligomerization resulting in apoptotic cell death in cancer cells[134]. Inhibiting apoptosis via p53-associated regulation of the BCL-2 family displays another level of complexity in inducing cell death of cancer cells.

7.2. BH3-mimetics as a therapeutic strategy to overcome apoptosis resistance

Due to the dysregulation and importance of BCL-2 family members for inhibiting apoptosis in cancer cells, attempts aimed at developing novel drugs that can inhibit anti-apoptotic BCL-2 proteins. Crystal structure analysis of BCL-xL revealed that the BH1-BH3 domains formed a hydrophobic groove[93]. Further studies demonstrated that this BCL-xL hydrophobic groove could bind to a BAK BH3 peptide indicating the ability to design small molecules that could bind to BCL-xL and inhibit its anti-apoptotic function[94]. Indeed, numerous small molecule BH3-mimetics have been identified or designed to bind to this

BH3 binding pocket with the potential to block BCL-2/xL binding to pro-apoptotic BCL-2 proteins. The BH3 mimetics have demonstrated diverse binding specificity and efficacy in inducing apoptosis (Figure 3)[135-137].

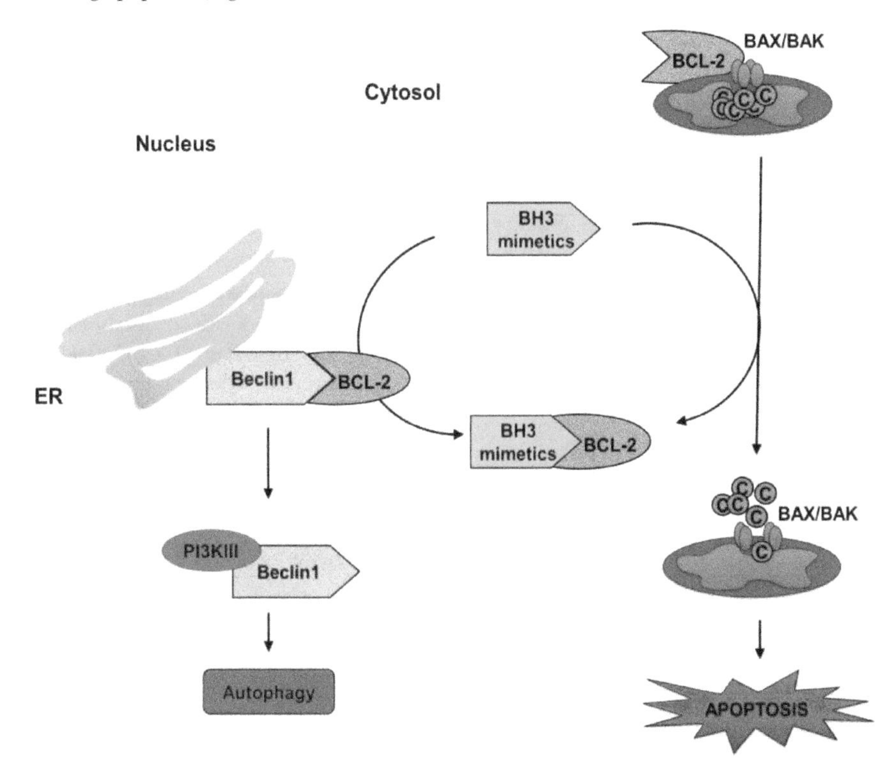

Figure 3. BH3 mimetics inhibit anti-apoptotic BCL-2 proteins therefore inducing both apoptosis and autophagy. BH3 mimetics are designed to bind to anti-apoptotic BCL-2 proteins and induce apoptosis. BH3 mimetics also induce autophagy-associated cell death by preventing BCL-2 proteins from binding to the autophagy activating protein, Beclin1.

One of the first small molecules developed via *in silico* screens was HA14-1[136]. HA14-1 was initially demonstrated to induce the activation of Apaf-1 and caspases in human acute myeloid leukemia cells. HA14-1 was subsequently found to prevent BCL-2 binding to BAK[138]. In addition, treatment with HA14-1 caused cytosolic Ca(2+) increase, change in mitochondrial membrane potential, BAX translocation, and reactive oxygen species (ROS) generation prior to cytochrome c release[139]. Obatoclax (GX15-070MS) was one of the first pan anti-apoptotic BCL-2 protein inhibitors capable of inhibiting BCL-2, BCL-XL, and MCL-1[140]. Clinical trials using obatoclax treatment have demonstrated success across many cancer types both independently as well as in combined therapies. Representative clinical trials are listed in Table 2.

Drug	Cancer type(s)	Clinical Trial	Co-therapy	Outcome
AT-101/ Gossypol	SCLC	Phase 2	None	Not active in patients with recurrent chemosensitive SCLC.[226]
	NSCLC	Phase 2	Docetaxel	AT-101 plus docetaxel was well tolerated with an adverse event profile indistinguishable from the base docetaxel regimen.[227]
	SCLC	Phase 1/2	Topotecan	Relapsed progression - 17.4 weeks, refractory progression - 11.7 weeks.[228]
	CRPC	Phase 1/2	None	Evidence of single-agent clinical activity was observed with prostate-specific antigen declines in some patients.[229]
	Metastatic Breast Cancer	Phase 1/2	None	Gossypol appears to affect the expression of Rb protein and cyclin D1; negligible antitumor activity against anthracycline and taxane refractory metastatic breast cancer.[230]
ABT-263/ Navitoclax	CLL	Phase 1	None	Low MCL1 expression and high BIM:MCL1 or BIM:BCL-2 ratios in leukemic cells correlated with response.[143]
	SCLC	Phase 1	None	Changes in a surrogate marker of BCL-2 amplification (pro-gastrin releasing peptide) correlated with changes in tumor volume.[144]
	Lymphoma	Phase 1	None	Navitoclax has a novel mechanism of peripheral thrombocytopenia and T-cell lymphopenia, attributable to high-affinity inhibition of BCL-XL and BCL-2, respectively.[231]
GX15-070MS/ Obatoclax mesylate	Leukemia	Phase I	None	Well tolerated and these results support its further investigation in patients with leukemia and myelodysplasia.[232]
	Solid tumors	Phase I	Topotecan	Safe and well tolerated when given in combination with topotecan.[233]
	CLL	Phase I	None	Activation of Bax and Bak was demonstrated in peripheral blood mononuclear cells, and apoptosis induction was related to obatoclax exposure, as monitored by the plasma concentration of oligonucleosomal DNA/histone complexes.[234]

Drug	Cancer type(s)	Clinical Trial	Co-therapy	Outcome
Oblimersen (Genasense)	Breast Cancer	Phase I	TAC	Two of 13 patients showed a decrease of BCL-2 transcripts after 4 days of treatment with oblimersen.[235]
	CRPC	Phase II	Docetaxel	The primary end points of the study were not met: PSA response rate >30% and a major toxic event rate <45% were not observed with docetaxel-oblimersen.[236]
	Breast Cancer	Phase II	TAC	Oblimersen up to a dose of 7 mg/kg/day administered as a 24-h infusion on days 1-7 can be safely administered in combination with standard TAC on day 5.[196]
	HRPC	Phase II	Docetaxel	Oblimersen combined with docetaxel is an active combination demonstrating both an encouraging response rate and an overall median survival. [237[

HRPC – Hormone Refractory Prostate Cancer; Chronic lymphocytic leukemia – CLL; Small Cell Lung Cancer- SCLC; Non-small cell lung cancer – NSCLC; Castrate-resistant prostate cancer – CRPC; TAC – docetaxel, adriamycin and cyclophosphamide.

Table 2. Selective list of published BH3 mimetics clinical trials with and without combination therapy.

Using nuclear magnetic resonance-based screening and structure-based design, the BH3 mimetic, ABT-737, was developed and shown to possess greater affinity and ability to inhibit BCL-2, BCL-xL and BCL-w, than MCL-1[141]. ABT-737 was initially developed by screening a library of BH-3 like analogues with high binding efficiency to the hydrophobic groove of BCL-xL. ABT-737 has been shown to synergistically enhance cell death in combined treatments with chemotherapeutics and radiation[141]. An oral form of ABT-737, called ABT-263 (Navitoclax), has also been developed and is also undergoing clinical trials for lymphoma, leukemia, and small cell lung cancer[142-144].

The BH3 mimetic (-)-gossypol is a natural polyphenol purified from the cottonseed. We previously demonstrated that the (-)-gossypol significantly enhances the antitumor activity of docetaxel chemotherapy in hormone-refractory prostate cancer patients with BCL-2/BCL-xL/MCL-1 overexpression[145]. Mechanistically, we demonstrated that (-)-gossypol blocked the interactions of BCL-2/Bcl-xL with Bax or Bad in cancer cells. (-)-Gossypol (AT-101) is the first BCL-2/BCL-xL inhibitor entered clinical trial and is now in Phase IIb clinical trials for hormone-refractory prostate cancer and many other types of cancer at multiple centers in the United States. In addition, more potent and less toxic gossypol derivatives, such as Apogossypolone and TW-37, are being developed[146-148].

BH3 mimetics are designed to inhibit anti-apoptotic BCL-2 proteins and demonstrate significant therapeutic potential in clinical trials. Interestingly, BH3 mimetics induced toxicity independent of Bax/Bak suggesting the existence of an alternative route of cell death induction[149]. BCL-2 has also been linked to a non-apoptotic cell death mechanism associated with autophagy, usually known as a cell survival mechanism[150]. It was later determined that BCL-2 and BCL-xL can bind the BH3 domain of tumor suppressor Beclin1 (BECN1) and inhibit autophagy (Figure 3)[151, 152]. This discovery revealed a new role for anti-apoptotic BCL-2 protein family as anti-autophagic proteins. The following sections will discuss autophagy and the role played by the BCL-2:Beclin 1 interaction for inducing/inhibiting autophagy, and the mechanism of novel therapies, such as BH3 mimetics, aimed at disrupting the interaction in order to induce autophagy-associated cell death.

8. Autophagy and autophagic cell death: background

Autophagy is a highly regulated catabolic process that functions as a cell survival mechanism activated upon cellular stresses such as nutrient deprivation, starvation, hypoxia and chemo/radiotherapy[153]. There are three primary types of autophagy, chaperone-mediate autophagy, microautophagy and macroautophagy[154]. This chapter will focus on macroautophagy, referred as autophagy further in the text. Activation of autophagy induces the formation of autophagosomes that engulf damaged organelles or particles. Eventually, the autophagsome fuses with the lysosome and degrades its interiors to provide cells the nutrients such as amino acids or fatty acids necessary for cell metabolism[155]. Defective autophagy machinery can lead to diseases such as neurodegenerative, liver, cardiac, and muscle diseases, as well as a variety of cancers. Recent studies have reported that apoptosis-resistant cancer cells can avoid chemo/radiotherapeutic-induced cell death by activating autophagy [156-159]. Furthermore, apoptosis-associated proteins, such as NF-κB, p53, UVRAG and the above-discussed BCL-2, have been shown to play dual regulatory roles in both apoptosis and autophagy [160-162]. Paradoxically, activation of autophagy upon drug treatments can induce cell death independent of or in parallel with apoptosis and necrosis.[163]. Therefore, researchers are actively developing novel cancer therapies that aim to promote cell death by modulating autophagy pathways.

8.1. Autophagy pathways

Autophagy can be activated by a variety of stimuli and signaling pathways. The classical induction of autophagy occurs upon nutrient deprivation; however, autophagy can also be induced by other factors, such as hypoxia, cytokines, hormones, genotoxic stress, p53 activation, and chemo/radiotherapy. Autophagy has also been attributed to tumor suppression. This was first demonstrated in mice with allelic loss of Beclin1, a key protein involved in inducing autophagy. Complete loss of the Beclin1 resulted in death during early embryogenesis whereas heterozygous loss of Beclin1 resulted in formations of spontaneous tumors [164, 165]. Autophagy involves a conserved family of proteins known as the autophagy-related gene families (ATGs). The canonical autophagy pathway in mammals occurs in a series of stages: initiation, nucleation, elongation, and degradation. All stages are regulated by a core molecular machinery (Figure 4).

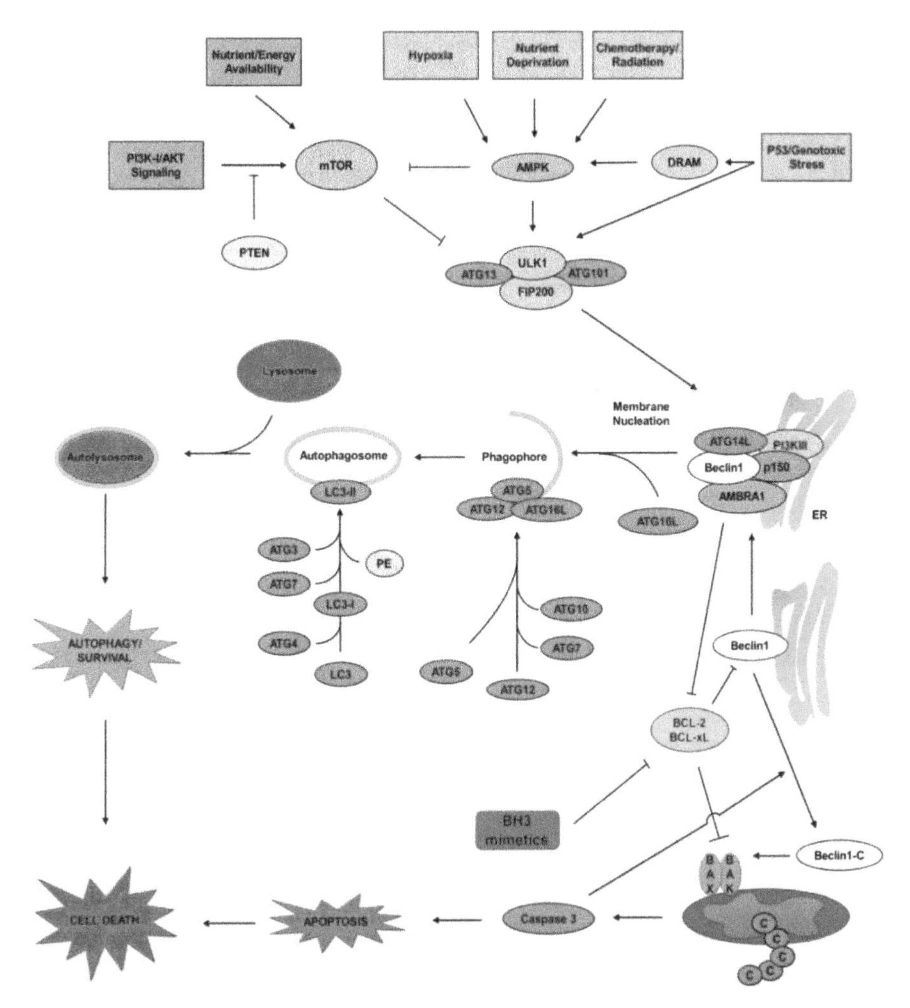

Figure 4. Cross-talk between apoptosis and autophagy. Autophagy takes place in a series of stages; initiation, nucleation, elongation, and degradation. Autophagy can be activated by a variety of stimuli and signaling pathways, including nutrient deprivation, hypoxia, p53 genotoxic stress, suppression of mTOR, or chemo/radiotherapy, followed by activation of AMPK. ULK1, ATG13, ATG101, FIP200 protein complex forms and mediates autophagy initiation. ATG13 mediates ULK1 phosphorylation of FIP200 and activates the ULK complex. Subsequently, the ULK complex localizes to the ER and initiates pre-autophagosome formation. The vesicle nucleation involves the core complex consisting of PI3KIII, p150, ATG14L, Beclin1 and AMBRA1. ATG14L induces a translocation of the PI3KIII complex to the site of autophagosome formation and initiates the formation of the phagophore. Phagophore elongation into an autophagosome requires ATG12, LC3-I, and two ubiquitin-like protein conjugation systems. The first system involves ATG7 and ATG10 conjugation of ATG12 into the ATG16L-ATG12-ATG5 complex. The second conjugation system involves LC3-I modification by ATG7 and ATG3 into LC3-II and inserts

into the autophagosome membrane. Finally, the autophagosome fuses with the lysosome and contents within the autophagosome are degraded. Beclin1 can interact with autophagy machinery at the ER and induce autophagy. In addition, Beclin1 can bind to anti-apoptotic BCL-2 family of proteins, preventing BCL-2 binding to BAX or BAK monomers, therefore inducing apoptosis. Beclin1 can also be cleaved by caspase 3 to form Beclin1-C which inhibits autophagy. However, Beclin1-C can induce apoptosis by localizing to the mitochondria facilitiating the release of apoptotic factors.

A protein complex consisting of unc-51-like kinase 1(ULK1, homolog of yeast ATG1), ATG13, ATG101 and a scaffolding protein FIP200 (ortholog of yeast ATG16) mediates autophagy initiation (Figure 4)[166]. In nutrient-rich environment, an upstream regulator called mammalian target of rapamycin (mTOR) phosphorylates ATG13 and ULK1 to inhibit the initiation of autophagy. Following starvation or cellular stress, mTOR is inhibited and dissociates from the ULK1 complex. Then, ATG13 mediates ULK1 to phosphorylate FIP200 and activates the ULK complex[167]. Subsequently, the ULK complex localizes to the endoplasmic reticulum (ER) and initiates pre-autophagosome formation [168]. The vesicle nucleation involves the core complex consisting of Class III phosphatidylinositol 3-kinase (PI3KIII/ homolog of yeast Vps34), p150 (Vps15), ATG14L, Beclin1 (ATG6), and activating molecule in Beclin 1-regulated autophagy (AMBRA1)[165, 169]. ATG14L induces a translocation of the PI3KIII complex to the site of autophagosome formation and initiates the formation of an isolated membrane, also known as the phagophore[170]. A recent study revealed that PI3KIII lipid kinase activity produces and accumulates phosphatidylinositol 3-phosphate (PI3P) at the ER to induce a high membrane curvature that attracts ATG14L binding. Bound to the ER, ATG14L produces more PI3Ps and recruits other parts of the core complex. Recruitment of these proteins induces phagophore elongation [171, 172]. Phagophore elongation into an autophagosome requires ATG12, ATG8/LC3-I, and two ubiquitin-like protein conjugation systems. The first system involves an E1-like ATG7 and an E2-like ATG10 conjugation of ATG12 to ATG12-ATG5 that interacts with ATG16L to form the ATG16L-ATG12-ATG5 complex[173-176]. The second conjugation system involves the cytosolic protein isoform known as the LC3-I (ATG8) to undergo modification by ATG7 and E2-like ATG3 into LC3-phosphatidylethanolamine (LC3-II), an important biomarker for autophagy[177, 178]. The ATG16L complex acts as an E3-like enzyme to promote lipidation of cytosolic LC3-I into LC3-II and correctly localizes LC3-II onto the autophagosome formation site to help form the membrane [179]. Finally, the autophagosome fuses with the lysosome and contents within the autophagosome are degraded. This final step requires the endosome marker, RAB7, and a lysosomal membrane protein, LAMP2, however, the exact mechanism involved in the fusion of autophagosome and lysosomes is still unclear [180, 181].

8.2. Autophagy induction

Activation of autophagy is regulated by multiple molecular pathways depending upon the stimuli (Figure 4). As mentioned above, mTOR is activated under nutrient-rich environment thereby suppressing autophagy. Starvation of growth factors and certain amino acids represses class I PI3K signaling to promote cell survival via autophagy induction [182, 183]. PI3KI forms the substrate PI3P which leads to activation of the PKB/AKT protein that inhibits a heterodimer complex involving the tuberous sclerosis complexes 1 and 2 (TSC1 and TSC2). The TSC2 protein suppresses mTOR activity via activation of a Ras family small GTPase called Ras homolog enriched in brain (Rheb) [184]. Tumor suppressor phosphatase

and tensin homolog (PTEN) dephosphorylates the PI3K product PI3P, thereby suppressing AKT signaling. Loss of PTEN occurs in multiple cancers including brain, breast, and prostate cancer [185]. Additional aberrant signaling of PI3KI can result in cancers that exhibit mutated amplification of upstream receptor tyrosine kinase, such as HER2 in gastric cancer or PDGFR and EGFR in glioblastoma [186, 187]. Under metabolic stress, such as high AMP level, hypoxia and cytosolic calcium level increase, AMP-activated protein kinase (AMPK) can mediate autophagy by negatively regulating mTOR and inducing the dephosphorylation of ATG13 and ULK1 [188-190]. Alternatively, AMPK has been found to activate autophagy by direction phosphorylation of ULK1 [191].

The tumor suppressor protein p53 plays a more complicated role and can induce as well as inhibit autophagy, based upon subcellular location and cellular context. Upon exposure to DNA-damaging agents, nuclear p53 can induce autophagy by transcriptionally activating damage-regulated autophagy modulator (DRAM)[192]. DRAM activates target proteins Sestrin1 and Sestrin2, which subsequently activate AMPK thereby inhibiting mTOR and inducing autophagy [193]. In addition, nuclear p53 can up-regulate ULK1 transcriptionally and directly activate autophagy[194]. Cytoplasmic p53 has the opposite effect and can actually inhibit autophagy[195]. High mobility group box 1 (HMGB1) is a Beclin1-interacting accessory protein that assists in autophagy activation. p53 has been discovered to form a complex with HMGB1 in the cytoplasm resulting in the inhibition of autophagy and induction of cell death (22345153)[87]. Loss or knockdown of p53 increases the binding of HMGB1 to Beclin1 and mediates cytosolic localization of the complex to the ER [87]. Subsequently, HMGB1 mediates the Beclin1-PI3KIII complex formation and initiates autophagosome production.

8.3. Beclin1:BCL2 interaction regulates autophagy/apoptosis switch

As discussed above, Beclin1 is a critical inducer of autophagy. Interestingly, Beclin1 is also a BH3-only protein and therefore interacts with anti-apoptotic BCL-2 family members via its BH3 domain [151, 196]. BCL-2 binding of Beclin1 at the ER prevents Beclin1 from assembling the pre-autophagosomal structure mediated by the Beclin1/PI3KIII complex (Figure 3 and 4)[196, 197]. Therefore, BCL-2 anti-apoptotic proteins have dual pro-survival roles by preventing both apoptosis and autophagy-associated cell death that makes these proteins ideal chemotherapeutic targets.

The expression of BCL-2 and/or Beclin1 is critical for regulating the switch between autophagy and apoptosis. Down-regulation of Beclin1 also contributes to tumorigenesis, evident in hepatocellular carcinoma, brain, colorectal, and gastric cancer [198-200]. Low expression of Beclin1 results in insufficient removal of damaged organelles. Deficient Beclin1 causes cell transformation through the accumulation of reactive oxygen species and genotoxic stress, [165]. Furthermore, it was shown that inhibiting BCL-2 in breast cancer cells via siRNA knockdown did not induce apoptosis as expected but observed a form of autophagic cell death [201]. The autophagic cell death was the result of combinatorial treatment with doxorubicin that lead to increased expression of Beclin1[201]. Therefore, maintaining adequate levels of Beclin1 is important to override BCL-2 inhibition of autophagy-related cell death. Evidently, BCL-2 and Beclin1 expressions are important determinants for identifying the proper chemotherapy or combination treatments that would provide the greatest therapeutic benefit.

Autophagy regulatory proteins can promote or inhibit the BCL-2:Beclin1 interactions. As previously mentioned, AMBRA1 is a key regulator in initiating autophagy by binding to Beclin1. In presence of autophagic stimulus, ULK1 phosphorylates AMBRA1, which results in AMBRA1 dissociation from the Dynein motor complex [202]. After dissociation, AMBRA1 translocates to the ER, binds to Beclin1 in the autophagy initiation complex and results in the induction of autophagy [202]. Moreover, a recent study demonstrated that BCL-2 localized to the mitochondria can also bind AMBRA1whereas ER-localized BCL-2 does not [203]. The BCL-2:AMBRA1 interaction at the mitochondria is down-regulated during autophagy and apoptosis. Therefore, BCL-2 can regulate Beclin1-induced autophagy by directly binding to Beclin1, as well as by sequestering AMBRA1, the activator of Beclin1 at the mitochondrion [203].

BCL-2/Beclin1 complex can be disrupted by otherBCL-2 and Beclin1 binding partners. As discussed above, HMGB1 can bind to Beclin1 and initiate autophagy. Inhibition of HMGB1 decreases autophagy and increases apoptosis[204]. For example, a study has shown that deletion or deactivation of HMGB1 in mouse embryonic fibroblasts reduces LC3-I expression. In response to starvation, cells lacking HMGB1 cannot initiate autophagy and undergo apoptotic cell death[205]. Additionally, HMGB1 bound to Beclin1 has also been found to induce the phosphorylation of BCL-2 which disrupts the BCL-2:Beclin1 complex.

Autophagy can also be inhibited by Beclin1 cleavage. Chemotherapy-induced and mitochondria-mediated apoptosis was shown to induce Beclin1 cleavage by caspase 8 to form Beclin1-C. This event renders defective Beclin1 activity and autophagy pathway [206]. Furthermore, the C-terminus of cleaved Beclin1 can acquire pro-apoptotic ability by translocation to the mitochondria and inducing release of apoptotic factors [207]. This demonstrates a novel therapeutic approach to induce apoptosis by inhibiting autophagy.

8.4. Inducing autophagy-associated cell death using BH3 mimetics

This chapter has previously discussed the therapeutic benefits of using BH3 mimetics to induce apoptosis by preventing anti-apoptotic BCL-2 proteins from binding to pro-apoptotic proteins, BAX/BAK. Upon the discovery that Beclin1 was a novel autophagic BH3-only protein, BH3 mimetics have been utilized to induce autophagy[208]. ABT737 was the first BH3 mimetic reported to induce apoptosis and autophagy by inhibiting anti-apoptotic action of BCL-2 or BCL-xL[208]. At first the findings were counterintuitive; how could a drug induce both apoptotic cell death and autophagic cell survival? As discussed above, BH3 mimetics appeared to kill cells in a BAK/BAX-independent manner suggesting that apoptotic cell death was not the only mechanism for BH3 mimetic-induced cell death[149]. It was later determined that BH3 mimetics could induce autophagy-associated cell death, especially in apoptosis-resistant cells [209].

We recently investigated the effect of the natural BH3-mimetic (-)-gossypol in apoptosis-resistant prostate cancer cells with high levels of BCL-2 versus prostate cancer cells with low BCL-2 expression[210]. (-)-Gossypol induced similar levels of total cell death in both prostate cancer cell lines. However, the dominant mode of cell death depended upon the expression of the anti-apoptotic BCL-2 family of proteins[210]. BH3 mimetics induced apoptotic cell death in prostate cancer cells with low BCL-2 expression. Conversely, prostate cancer cells with high BCL-2 expression died via modulation of the autophagy pathway [210]. Furthermore, overexpressing BCL-2 decreased the level of (-)-gossypol-induced autophagy, possibly due to the stoichiometric abundance of BCL-2 sequestering Beclin1 and inhibiting autophagy induction. The data demonstrate that BH3 mimetics can be utilized to kill cells with both high and low BCL-2, therefore, enhancing the ability to overcome chemo/radioresistance.

BH3 mimetics induce autophagy by disrupting the BCL-2:Beclin1 inhibitory complex as well as additional autophagy pathways. BH3 mimetics, ABT-737 and HA14-1, also stimulate other pro-autophagic pathways and hence activate the nutrient sensors Sirtuin1 and AMPK, inhibit mTOR, deplete cytoplasmic p53, and trigger the IKK Kinase[211]. Activation of autophagy was independent of reduced oxidative phosphorylation or reduced cellular ATP concentrations. Furthermore, induction of autophagy by ABT-737 and HA14-1 was completely inhibited by knockdown of Beclin1 or PI3KIII. This suggests that BH3 mimetics can interfere with multiple pathways, eliciting a coordinated effort to induce autophagy-associated cell death.

9. The role of autophagy in therapy resistance

A number of therapeutic strategies have been developed to target autophagy in cancer cells. Similarly with apoptosis-resistance, autophagy-associated resistance to chemotherapy has become a challenging variable in the successful treatment of patients. For example, in human lung cancer cells treated with EGFR tyrosine kinase inhibitors (TKI), gefitinib and erlotinib, autophagy contributed to cell survival[212]. Inhibition of EGFR suppresses PI3KI activity and results in downstream activation of the ULK complex[212]. Other studies have shown that autophagy contributes to chemotherapy resistance through its cytoprotective mechanism. For example, chronic myeloid leukemia treated with imatinib, glioblastoma multiforme treated with temozolomide, colorectal cancer treated with 5-FU, and breast cancer treated with both tamoxifen and trastuzumab have all shown resistance that is associated with increased autophagy[213]. Recent studies have shown that cytotoxic agents and starvation may play a role in activating autophagy via HMGB1[214]. Increased expression of HMGB1 during treatment with doxorubicin, cisplatin, and methotrexate in osteosarcoma patients has been found to facilitate chemotherapy resistance by promoting the formation Beclin1/PI3KIII complex. In addition, HMGB1 also antagonizes drug-induced cell death in leukemia, colon cancer, and prostate cancer by up-regulating autophagy but the exact mechanism remains unclear[214].

Not only does autophagy contribute to chemotherapy resistance, it also plays a role in radiotherapy resistance. Investigators exposed radioresistant MDA-MB-231 cells to ionizing radiation at different doses and found increasing levels of LC3-II, a hallmark of autophagy activation. This indicates that activation of autophagy may protect these cells from radiation-induced cell death[215]. In addition, researchers found upregulation of autophagy in radiosensitive HBL-100 cells after inhibition of mTOR by rapamycin. In further experiments, inhibition of autophagy by 3-methyladenine (3-MA) resulted in reduced cell survival and displayed a radiosensitizing effect[215]. From these experiments, researchers deduced that cancer cells use autophagy as an escape mechanism from apoptosis to overcome radiotherapeutic stress via degradation of IR-induced cellular damage.

10. Re-sensitization of cancer cells to treatment by autophagy inhibition

To counter autophagy in cancer resistance, novel cancer therapies uses target inhibition of autophagy for re-sensitizing cancer cells to drug treatments. Researchers have used autophagy inhibitors such as 3-MA, LY294002, wortmannin to inhibit the PI3K [158, 216]. 3-MA contributes to autophagy suppression by down-regulating the PI3KI/Akt/mTOR signaling pathway. Surprisingly, the autophagy inhibitor 3-MA has been found to induce autophagy and contribute to cell survival when used for a prolonged period[217]. This controversial phenomenon is most likely due to the dual effect of 3-MA on PI3KI and PI3KIII. 3-MA blocks Class I permanently, but only temporarily Class III PI3K. Thus, treatment with 3-MA should only be considered under specific conditions such as limited treatment periods. Other types of autophagy inhibitors include LC3 knockdown by siRNA, which decreased breast cancer resistance to trastuzumab and increased cell death in CML in combination with imatinib[156, 218]. Chloroquine (CQ) and Hydroxychloroquine (HCQ) are the most successful autophagy inhibitors that suppress the autophagic lysosomal protease activity to promote the accumulation of autophagic vacuoles that often leads to apoptotic and necrotic cell death[219, 220]. Phase I and II clinical trials are ongoing using HCQ or CQ in combination with treatment such as docetaxel in prostate cancer, tamoxifen in breast cancer, and gemcitabine in pancreatic cancer[221].

Inhibiting autophagy poses another potential problem since anti-autophagic therapeutic drugs reduce tumor-specific immune response thereby limiting the therapeutic success[222]. Activated autophagy in glioblastoma cells treated with EGF toxin has been found to release HMGB1 that binds to and activates Toll-like receptor 4 (TLR4). Activated TLR4 increases T-cell mediated anti-tumor response to eliminate the malignant cells[223]. Deactivating autophagy decreases the release of HMGB1, leaves tumor cells unattended by the host immune system, and results in increased resistance[224]. Although inhibiting autophagy is effective, researchers must take its adverse side effects into consideration.

11. Concluding remarks

As this chapter has outlined, chemo/radioresistance is a key contributor to decreased patient survival. In order to develop more effective cancer therapies and improve treatment outcome, more research is required to delineate this complicated biological mechanism. Furthermore, the ability of cancer cells to acquire heterogeneous genetic and epigenetic alterations across tumors elicits deregulation of cell death-associated signaling pathways in a variety of ways. Cancer cells are smart to quickly figure out ways to overcome a treatment that targets any particular cellular signaling pathway. Therefore, designing novel drugs and enhancing therapeutic strategies must simultaneously target multiple pathways and mechanisms. Using IAP antagonists that target multiple cell survival pathways, as well as BH3 mimetics that can overcome anti-apoptotic BCL-2 proteins to induce both apoptosis and autophagy-related cell death can improve survival and quality of life for cancer patients. The complexity of tumor biology and drug resistance suggests that we need to design treatment strategy based on the genetic/signaling profiles of the patient in order to provide the safest and most effective cancer therapies tailored to a particular patient, the ultimate goal of the personalized medicine.

Author details

Rebecca T. Marquez, Bryan W. Tsao and Nicholas F. Faust
Department of Molecular Biosciences, University of Kansas, Lawrence, Kansas, USA

Liang Xu *
Department of Molecular Biosciences, University of Kansas, Lawrence, Kansas, USA
Departments of Radiation Oncology and Urology,
University of Kansas Medical School, Kansas City, Kansas, USA
University of Kansas Cancer Center, Kansas City, Kansas, USA

Acknowledgement

This study was supported in part by NIH grants R01 CA121830(S1) and R01 CA134655 (LX) a pilot grant from NIH COBRE CCET grant (8P30GM103495), Bridging Grant from the Kansas IDeA Network of Biomedical Research Excellence (K-INBRE) (P20 GM103418), and by the Kansas Bioscience Authority Rising Star Award (LX). RTM is supported in part by NIH K-INBRE Post-Doctoral Award. Its contents are solely the responsibility of the authors and do not necessarily represent the official views of NIH.

* Corresponding Author

12. References

[1] Galluzzi L, Vitale I, Abrams JM, Alnemri ES, Baehrecke EH, Blagosklonny MV, et al. Molecular definitions of cell death subroutines: recommendations of the Nomenclature Committee on Cell Death 2012. Cell death and differentiation. 2012;19(1):107-20. Epub 2011/07/16.

[2] Fulda S, Galluzzi L, Kroemer G. Targeting mitochondria for cancer therapy. Nature reviews Drug discovery. 2010;9(6):447-64. Epub 2010/05/15.

[3] Schimmer AD. Inhibitor of apoptosis proteins: translating basic knowledge into clinical practice. Cancer research. 2004;64(20):7183-90. Epub 2004/10/20.

[4] Srinivasula SM, Ashwell JD. IAPs: what's in a name? Molecular cell. 2008;30(2):123-35. Epub 2008/04/29.

[5] Deveraux QL, Stennicke HR, Salvesen GS, Reed JC. Endogenous inhibitors of caspases. J Clin Immunol. 1999;19(6):388-98. Epub 2000/01/14.

[6] Fulda S, Vucic D. Targeting IAP proteins for therapeutic intervention in cancer. Nature reviews Drug discovery. 2012;11(2):109-24. Epub 2012/02/02.

[7] Takahashi R, Deveraux Q, Tamm I, Welsh K, Assa-Munt N, Salvesen GS, et al. A single BIR domain of XIAP sufficient for inhibiting caspases. The Journal of biological chemistry. 1998;273(14):7787-90. Epub 1998/05/09.

[8] Dai Y, Lawrence T, Xu L. Overcoming cancer therapy resistance by targeting inhibitors of apoptosis proteins and nuclear factor-kappa B. American journal of translational research. 2009;1(eb15eeff-35e4-8951-208a-fb085ebac5df):1-16.

[9] Vucic D, Dixit VM, Wertz IE. Ubiquitylation in apoptosis: a post-translational modification at the edge of life and death. Nature reviews Molecular cell biology. 2011;12(7):439-52. Epub 2011/06/24.

[10] Deveraux QL, Takahashi R, Salvesen GS, Reed JC. X-linked IAP is a direct inhibitor of cell-death proteases. Nature. 1997;388(6639):300-4. Epub 1997/07/17.

[11] Wilkinson JC, Cepero E, Boise LH, Duckett CS. Upstream regulatory role for XIAP in receptor-mediated apoptosis. Mol Cell Biol. 2004;24(16):7003-14. Epub 2004/07/30.

[12] Fong WG, Liston P, Rajcan-Separovic E, St Jean M, Craig C, Korneluk RG. Expression and genetic analysis of XIAP-associated factor 1 (XAF1) in cancer cell lines. Genomics. 2000;70(1):113-22. Epub 2000/11/23.

[13] Muris JJ, Cillessen SA, Vos W, van Houdt IS, Kummer JA, van Krieken JH, et al. Immunohistochemical profiling of caspase signaling pathways predicts clinical response to chemotherapy in primary nodal diffuse large B-cell lymphomas. Blood. 2005;105(7):2916-23. Epub 2004/12/04.

[14] Yang XH, Feng ZE, Yan M, Hanada S, Zuo H, Yang CZ, et al. XIAP is a predictor of cisplatin-based chemotherapy response and prognosis for patients with advanced head and neck cancer. PloS one. 2012;7(3):e31601. Epub 2012/03/10.

[15] Hector S, Rehm M, Schmid J, Kehoe J, McCawley N, Dicker P, et al. Clinical application of a systems model of apoptosis execution for the prediction of colorectal cancer

therapy responses and personalisation of therapy. Gut. 2012;61(5):725-33. Epub 2011/11/16.

[16] Tamm I, Kornblau SM, Segall H, Krajewski S, Welsh K, Kitada S, et al. Expression and prognostic significance of IAP-family genes in human cancers and myeloid leukemias. Clinical cancer research : an official journal of the American Association for Cancer Research. 2000;6(5):1796-803. Epub 2000/05/18.

[17] Deveraux QL, Leo E, Stennicke HR, Welsh K, Salvesen GS, Reed JC. Cleavage of human inhibitor of apoptosis protein XIAP results in fragments with distinct specificities for caspases. The EMBO journal. 1999;18(19):5242-51. Epub 1999/10/03.

[18] Datta R, Oki E, Endo K, Biedermann V, Ren J, Kufe D. XIAP regulates DNA damage-induced apoptosis downstream of caspase-9 cleavage. The Journal of biological chemistry. 2000;275(41):31733-8. Epub 2000/08/10.

[19] Deveraux QL, Roy N, Stennicke HR, Van Arsdale T, Zhou Q, Srinivasula SM, et al. IAPs block apoptotic events induced by caspase-8 and cytochrome c by direct inhibition of distinct caspases. The EMBO journal. 1998;17(8):2215-23. Epub 1998/05/26.

[20] Srinivasula SM, Hegde R, Saleh A, Datta P, Shiozaki E, Chai J, et al. A conserved XIAP-interaction motif in caspase-9 and Smac/DIABLO regulates caspase activity and apoptosis. Nature. 2001;410(6824):112-6. Epub 2001/03/10.

[21] Shiozaki EN, Chai J, Rigotti DJ, Riedl SJ, Li P, Srinivasula SM, et al. Mechanism of XIAP-mediated inhibition of caspase-9. Molecular cell. 2003;11(2):519-27. Epub 2003/03/07.

[22] Huang Y, Park YC, Rich RL, Segal D, Myszka DG, Wu H. Structural basis of caspase inhibition by XIAP: differential roles of the linker versus the BIR domain. Cell. 2001;104(5):781-90. Epub 2001/03/21.

[23] Suzuki Y, Nakabayashi Y, Takahashi R. Ubiquitin-protein ligase activity of X-linked inhibitor of apoptosis protein promotes proteasomal degradation of caspase-3 and enhances its anti-apoptotic effect in Fas-induced cell death. Proceedings of the National Academy of Sciences of the United States of America. 2001;98(15):8662-7. Epub 2001/07/12.

[24] Pop C, Salvesen GS. Human caspases: activation, specificity, and regulation. The Journal of biological chemistry. 2009;284(33):21777-81. Epub 2009/05/29.

[25] Magne N, Toillon RA, Bottero V, Didelot C, Houtte PV, Gerard JP, et al. NF-kappaB modulation and ionizing radiation: mechanisms and future directions for cancer treatment. Cancer Lett. 2006;231(2):158-68. Epub 2006/01/10.

[26] Voboril R, Weberova-Voborilova J. Constitutive NF-kappaB activity in colorectal cancer cells: impact on radiation-induced NF-kappaB activity, radiosensitivity, and apoptosis. Neoplasma. 2006;53(6):518-23. Epub 2006/12/15.

[27] Rho HS, Kim SH, Lee CE. Mechanism of NF-kappaB activation induced by gamma-irradiation in B lymphoma cells : role of Ras. J Toxicol Environ Health A. 2005;68(23-24):2019-31. Epub 2005/12/06.

[28] Scheidereit C. IkappaB kinase complexes: gateways to NF-kappaB activation and transcription. Oncogene. 2006;25(51):6685-705. Epub 2006/10/31.

[29] Darding M, Meier P. IAPs: guardians of RIPK1. Cell death and differentiation. 2012;19(1):58-66. Epub 2011/11/19.

[30] Micheau O, Tschopp J. Induction of TNF receptor I-mediated apoptosis via two sequential signaling complexes. Cell. 2003;114(2):181-90. Epub 2003/07/31.

[31] Rothe M, Pan MG, Henzel WJ, Ayres TM, Goeddel DV. The TNFR2-TRAF signaling complex contains two novel proteins related to baculoviral inhibitor of apoptosis proteins. Cell. 1995;83(7):1243-52. Epub 1995/12/29.

[32] Varfolomeev E, Goncharov T, Fedorova AV, Dynek JN, Zobel K, Deshayes K, et al. c-IAP1 and c-IAP2 are critical mediators of tumor necrosis factor alpha (TNFalpha)-induced NF-kappaB activation. J Biol Chem. 2008;283(36):24295-9. Epub 2008/07/16.

[33] Bertrand MJ, Milutinovic S, Dickson KM, Ho WC, Boudreault A, Durkin J, et al. cIAP1 and cIAP2 facilitate cancer cell survival by functioning as E3 ligases that promote RIP1 ubiquitination. Molecular cell. 2008;30(6):689-700. Epub 2008/06/24.

[34] Mahoney DJ, Cheung HH, Mrad RL, Plenchette S, Simard C, Enwere E, et al. Both cIAP1 and cIAP2 regulate TNFalpha-mediated NF-kappaB activation. Proceedings of the National Academy of Sciences of the United States of America. 2008;105(33):11778-83. Epub 2008/08/14.

[35] Varfolomeev E, Blankenship JW, Wayson SM, Fedorova AV, Kayagaki N, Garg P, et al. IAP antagonists induce autoubiquitination of c-IAPs, NF-kappaB activation, and TNFalpha-dependent apoptosis. Cell. 2007;131(4):669-81. Epub 2007/11/21.

[36] Demchenko YN, Glebov OK, Zingone A, Keats JJ, Bergsagel PL, Kuehl WM. Classical and/or alternative NF-kappaB pathway activation in multiple myeloma. Blood. 2010;115(17):3541-52. Epub 2010/01/08.

[37] Annunziata CM, Davis RE, Demchenko Y, Bellamy W, Gabrea A, Zhan F, et al. Frequent engagement of the classical and alternative NF-kappaB pathways by diverse genetic abnormalities in multiple myeloma. Cancer cell. 2007;12(2):115-30. Epub 2007/08/19.

[38] Vince JE, Wong WW, Khan N, Feltham R, Chau D, Ahmed AU, et al. IAP antagonists target cIAP1 to induce TNFalpha-dependent apoptosis. Cell. 2007;131(4):682-93. Epub 2007/11/21.

[39] Keats JJ, Fonseca R, Chesi M, Schop R, Baker A, Chng WJ, et al. Promiscuous mutations activate the noncanonical NF-kappaB pathway in multiple myeloma. Cancer cell. 2007;12(2):131-44. Epub 2007/08/19.

[40] He JQ, Zarnegar B, Oganesyan G, Saha SK, Yamazaki S, Doyle SE, et al. Rescue of TRAF3-null mice by p100 NF-kappa B deficiency. The Journal of experimental medicine. 2006;203(11):2413-8. Epub 2006/10/04.

[41] Grech AP, Amesbury M, Chan T, Gardam S, Basten A, Brink R. TRAF2 differentially regulates the canonical and noncanonical pathways of NF-kappaB activation in mature B cells. Immunity. 2004;21(5):629-42. Epub 2004/11/13.

[42] Varfolomeev E, Goncharov T, Maecker H, Zobel K, Komuves LG, Deshayes K, et al. Cellular Inhibitors of Apoptosis Are Global Regulators of NF-kappaB and MAPK Activation by Members of the TNF Family of Receptors. Sci Signal. 2012;5(216):ra22. Epub 2012/03/22.

[43] Wu G, Chai J, Suber TL, Wu JW, Du C, Wang X, et al. Structural basis of IAP recognition by Smac/DIABLO. Nature. 2000;408(6815):1008-12. Epub 2001/01/05.

[44] Chai J, Du C, Wu JW, Kyin S, Wang X, Shi Y. Structural and biochemical basis of apoptotic activation by Smac/DIABLO. Nature. 2000;406(6798):855-62. Epub 2000/09/06.

[45] Liu Z, Sun C, Olejniczak ET, Meadows RP, Betz SF, Oost T, et al. Structural basis for binding of Smac/DIABLO to the XIAP BIR3 domain. Nature. 2000;408(6815):1004-8. Epub 2001/01/05.

[46] Salvesen GS, Duckett CS. IAP proteins: blocking the road to death's door. Nature reviews Molecular cell biology. 2002;3(6):401-10. Epub 2002/06/04.

[47] Vucic D, Fairbrother WJ. The inhibitor of apoptosis proteins as therapeutic targets in cancer. Clinical cancer research : an official journal of the American Association for Cancer Research. 2007;13(20):5995-6000. Epub 2007/10/20.

[48] Imoto I, Yang ZQ, Pimkhaokham A, Tsuda H, Shimada Y, Imamura M, et al. Identification of cIAP1 as a candidate target gene within an amplicon at 11q22 in esophageal squamous cell carcinomas. Cancer research. 2001;61(18):6629-34. Epub 2001/09/18.

[49] Zender L, Spector MS, Xue W, Flemming P, Cordon-Cardo C, Silke J, et al. Identification and validation of oncogenes in liver cancer using an integrative oncogenomic approach. Cell. 2006;125(7):1253-67. Epub 2006/07/04.

[50] Imoto I, Tsuda H, Hirasawa A, Miura M, Sakamoto M, Hirohashi S, et al. Expression of cIAP1, a target for 11q22 amplification, correlates with resistance of cervical cancers to radiotherapy. Cancer research. 2002;62(17):4860-6. Epub 2002/09/05.

[51] Dai Z, Zhu WG, Morrison CD, Brena RM, Smiraglia DJ, Raval A, et al. A comprehensive search for DNA amplification in lung cancer identifies inhibitors of apoptosis cIAP1 and cIAP2 as candidate oncogenes. Human molecular genetics. 2003;12(7):791-801. Epub 2003/03/26.

[52] Akagi T, Motegi M, Tamura A, Suzuki R, Hosokawa Y, Suzuki H, et al. A novel gene, MALT1 at 18q21, is involved in t(11;18) (q21;q21) found in low-grade B-cell lymphoma of mucosa-associated lymphoid tissue. Oncogene. 1999;18(42):5785-94. Epub 1999/10/19.

[53] Zhou H, Du MQ, Dixit VM. Constitutive NF-kappaB activation by the t(11;18)(q21;q21) product in MALT lymphoma is linked to deregulated ubiquitin ligase activity. Cancer cell. 2005;7(5):425-31. Epub 2005/05/17.

[54] Morgan JA, Yin Y, Borowsky AD, Kuo F, Nourmand N, Koontz JI, et al. Breakpoints of the t(11;18)(q21;q21) in mucosa-associated lymphoid tissue (MALT) lymphoma lie within or near the previously undescribed gene MALT1 in chromosome 18. Cancer research. 1999;59(24):6205-13. Epub 2000/01/08.

[55] Varfolomeev E, Wayson SM, Dixit VM, Fairbrother WJ, Vucic D. The inhibitor of apoptosis protein fusion c-IAP2.MALT1 stimulates NF-kappaB activation independently of TRAF1 AND TRAF2. The Journal of biological chemistry. 2006;281(39):29022-9. Epub 2006/08/08.

[56] Ramp U, Krieg T, Caliskan E, Mahotka C, Ebert T, Willers R, et al. XIAP expression is an independent prognostic marker in clear-cell renal carcinomas. Hum Pathol. 2004;35(8):1022-8. Epub 2004/08/07.

[57] Li M, Song T, Yin ZF, Na YQ. XIAP as a prognostic marker of early recurrence of nonmuscular invasive bladder cancer. Chinese medical journal. 2007;120(6):469-73. Epub 2007/04/19.

[58] Xiang G, Wen X, Wang H, Chen K, Liu H. Expression of X-linked inhibitor of apoptosis protein in human colorectal cancer and its correlation with prognosis. Journal of surgical oncology. 2009;100(8):708-12. Epub 2009/09/25.

[59] Shi YH, Ding WX, Zhou J, He JY, Xu Y, Gambotto AA, et al. Expression of X-linked inhibitor-of-apoptosis protein in hepatocellular carcinoma promotes metastasis and tumor recurrence. Hepatology. 2008;48(2):497-507. Epub 2008/07/31.

[60] Zhang Y, Zhu J, Tang Y, Li F, Zhou H, Peng B, et al. X-linked inhibitor of apoptosis positive nuclear labeling: a new independent prognostic biomarker of breast invasive ductal carcinoma. Diagnostic pathology. 2011;6:49. Epub 2011/06/08.

[61] Seligson DB, Hongo F, Huerta-Yepez S, Mizutani Y, Miki T, Yu H, et al. Expression of X-linked inhibitor of apoptosis protein is a strong predictor of human prostate cancer recurrence. Clinical cancer research : an official journal of the American Association for Cancer Research. 2007;13(20):6056-63. Epub 2007/10/20.

[62] Yan H, Yu J, Wang R, Jiang S, Zhu K, Mu D, et al. Prognostic value of Smac expression in rectal cancer patients treated with neoadjuvant therapy. Med Oncol. 2012;29(1):168-73. Epub 2011/01/26.

[63] Sekimura A, Konishi A, Mizuno K, Kobayashi Y, Sasaki H, Yano M, et al. Expression of Smac/DIABLO is a novel prognostic marker in lung cancer. Oncol Rep. 2004;11(4):797-802. Epub 2004/03/11.

[64] Kempkensteffen C, Jager T, Bub J, Weikert S, Hinz S, Christoph F, et al. The equilibrium of XIAP and Smac/DIABLO expression is gradually deranged during the development and progression of testicular germ cell tumours. Int J Androl. 2007;30(5):476-83. Epub 2007/02/15.

[65] Bao ST, Gui SQ, Lin MS. Relationship between expression of Smac and Survivin and apoptosis of primary hepatocellular carcinoma. Hepatobiliary Pancreat Dis Int. 2006;5(4):580-3. Epub 2006/11/07.

[66] Yan Y, Mahotka C, Heikaus S, Shibata T, Wethkamp N, Liebmann J, et al. Disturbed balance of expression between XIAP and Smac/DIABLO during tumour progression in renal cell carcinomas. Br J Cancer. 2004;91(7):1349-57. Epub 2004/08/26.

[67] Kashkar H, Haefs C, Shin H, Hamilton-Dutoit SJ, Salvesen GS, Kronke M, et al. XIAP-mediated caspase inhibition in Hodgkin's lymphoma-derived B cells. The Journal of experimental medicine. 2003;198(2):341-7. Epub 2003/07/23.
[68] Kashkar H, Seeger JM, Hombach A, Deggerich A, Yazdanpanah B, Utermohlen O, et al. XIAP targeting sensitizes Hodgkin lymphoma cells for cytolytic T-cell attack. Blood. 2006;108(10):3434-40. Epub 2006/07/27.
[69] Fulda S, Wick W, Weller M, Debatin KM. Smac agonists sensitize for Apo2L/TRAIL- or anticancer drug-induced apoptosis and induce regression of malignant glioma in vivo. Nat Med. 2002;8(8):808-15. Epub 2002/07/16.
[70] Pardo OE, Lesay A, Arcaro A, Lopes R, Ng BL, Warne PH, et al. Fibroblast growth factor 2-mediated translational control of IAPs blocks mitochondrial release of Smac/DIABLO and apoptosis in small cell lung cancer cells. Mol Cell Biol. 2003;23(21):7600-10. Epub 2003/10/16.
[71] Li L, Thomas RM, Suzuki H, De Brabander JK, Wang X, Harran PG. A small molecule Smac mimic potentiates TRAIL- and TNFalpha-mediated cell death. Science. 2004;305(5689):1471-4. Epub 2004/09/09.
[72] Sun H, Nikolovska-Coleska Z, Yang CY, Xu L, Tomita Y, Krajewski K, et al. Structure-based design, synthesis, and evaluation of conformationally constrained mimetics of the second mitochondria-derived activator of caspase that target the X-linked inhibitor of apoptosis protein/caspase-9 interaction site. Journal of medicinal chemistry. 2004;47(17):4147-50. Epub 2004/08/06.
[73] Sun H, Nikolovska-Coleska Z, Yang CY, Xu L, Liu M, Tomita Y, et al. Structure-based design of potent, conformationally constrained Smac mimetics. J Am Chem Soc. 2004;126(51):16686-7. Epub 2004/12/23.
[74] Sun H, Nikolovska-Coleska Z, Lu J, Qiu S, Yang CY, Gao W, et al. Design, synthesis, and evaluation of a potent, cell-permeable, conformationally constrained second mitochondria derived activator of caspase (Smac) mimetic. Journal of medicinal chemistry. 2006;49(26):7916-20. Epub 2006/12/22.
[75] Dai Y, Liu M, Tang W, DeSano J, Burstein E, Davis M, et al. Molecularly targeted radiosensitization of human prostate cancer by modulating inhibitor of apoptosis. Clinical cancer research : an official journal of the American Association for Cancer Research. 2008;14(23):7701-10. Epub 2008/12/03.
[76] Dai Y, Liu M, Tang W, Li Y, Lian J, Lawrence TS, et al. A Smac-mimetic sensitizes prostate cancer cells to TRAIL-induced apoptosis via modulating both IAPs and NF-kappaB. BMC Cancer. 2009;9:392. Epub 2009/11/10.
[77] Schimmer AD, Welsh K, Pinilla C, Wang Z, Krajewska M, Bonneau MJ, et al. Small-molecule antagonists of apoptosis suppressor XIAP exhibit broad antitumor activity. Cancer cell. 2004;5(1):25-35. Epub 2004/01/30.
[78] Petersen SL, Wang L, Yalcin-Chin A, Li L, Peyton M, Minna J, et al. Autocrine TNFalpha signaling renders human cancer cells susceptible to Smac-mimetic-induced apoptosis. Cancer cell. 2007;12(5):445-56. Epub 2007/11/13.

[79] Ndubaku C, Varfolomeev E, Wang L, Zobel K, Lau K, Elliott LO, et al. Antagonism of c-IAP and XIAP proteins is required for efficient induction of cell death by small-molecule IAP antagonists. ACS chemical biology. 2009;4(7):557-66. Epub 2009/06/06.
[80] Nikolovska-Coleska Z, Xu L, Hu Z, Tomita Y, Li P, Roller PP, et al. Discovery of embelin as a cell-permeable, small-molecular weight inhibitor of XIAP through structure-based computational screening of a traditional herbal medicine three-dimensional structure database. Journal of medicinal chemistry. 2004;47(10):2430-40. Epub 2004/04/30.
[81] Dai Y, Desano J, Qu Y, Tang W, Meng Y, Lawrence TS, et al. Natural IAP inhibitor Embelin enhances therapeutic efficacy of ionizing radiation in prostate cancer. American journal of cancer research. 2011;1(2):128-43. Epub 2011/08/02.
[82] Zhang S, Ding F, Luo A, Chen A, Yu Z, Ren S, et al. XIAP is highly expressed in esophageal cancer and its downregulation by RNAi sensitizes esophageal carcinoma cell lines to chemotherapeutics. Cancer Biol Ther. 2007;6(6):973-80. Epub 2007/07/06.
[83] McManus DC, Lefebvre CA, Cherton-Horvat G, St-Jean M, Kandimalla ER, Agrawal S, et al. Loss of XIAP protein expression by RNAi and antisense approaches sensitizes cancer cells to functionally diverse chemotherapeutics. Oncogene. 2004;23(49):8105-17. Epub 2004/09/21.
[84] Cao C, Mu Y, Hallahan DE, Lu B. XIAP and survivin as therapeutic targets for radiation sensitization in preclinical models of lung cancer. Oncogene. 2004;23(42):7047-52. Epub 2004/07/20.
[85] Dean E, Jodrell D, Connolly K, Danson S, Jolivet J, Durkin J, et al. Phase I trial of AEG35156 administered as a 7-day and 3-day continuous intravenous infusion in patients with advanced refractory cancer. J Clin Oncol. 2009;27(10):1660-6. Epub 2009/02/25.
[86] Mahadevan D, Chalasani P, Rensvold D, Kurtin S, Pretzinger C, Jolivet J, et al. Phase I Trial of AEG35156 an Antisense Oligonucleotide to XIAP Plus Gemcitabine in Patients With Metastatic Pancreatic Ductal Adenocarcinoma. Am J Clin Oncol. 2012. Epub 2012/03/24.
[87] Schimmer AD, Estey EH, Borthakur G, Carter BZ, Schiller GJ, Tallman MS, et al. Phase I/II trial of AEG35156 X-linked inhibitor of apoptosis protein antisense oligonucleotide combined with idarubicin and cytarabine in patients with relapsed or primary refractory acute myeloid leukemia. J Clin Oncol. 2009;27(28):4741-6. Epub 2009/08/05.
[88] Karnak D, Xu L. Chemosensitization of prostate cancer by modulating Bcl-2 family proteins. Current drug targets. 2010;11(c06aabad-8fbd-e7ce-4422-fb085eb07a5c):699-1406.
[89] Adams JM, Cory S. The Bcl-2 apoptotic switch in cancer development and therapy. Oncogene. 2007;26(9):1324-37. Epub 2007/02/27.
[90] Kroemer G, Galluzzi L, Brenner C. Mitochondrial membrane permeabilization in cell death. Physiol Rev. 2007;87(1):99-163. Epub 2007/01/24.

[91] Youle RJ, Strasser A. The BCL-2 protein family: opposing activities that mediate cell death. Nature reviews Molecular cell biology. 2008;9(1):47-59. Epub 2007/12/22.
[92] Chan SL, Yu VC. Proteins of the bcl-2 family in apoptosis signalling: from mechanistic insights to therapeutic opportunities. Clinical and experimental pharmacology & physiology. 2004;31(3):119-28. Epub 2004/03/11.
[93] Muchmore SW, Sattler M, Liang H, Meadows RP, Harlan JE, Yoon HS, et al. X-ray and NMR structure of human Bcl-xL, an inhibitor of programmed cell death. Nature. 1996;381(6580):335-41. Epub 1996/05/23.
[94] Sattler M, Liang H, Nettesheim D, Meadows RP, Harlan JE, Eberstadt M, et al. Structure of Bcl-xL-Bak peptide complex: recognition between regulators of apoptosis. Science. 1997;275(5302):983-6. Epub 1997/02/14.
[95] Zhu W, Cowie A, Wasfy GW, Penn LZ, Leber B, Andrews DW. Bcl-2 mutants with restricted subcellular location reveal spatially distinct pathways for apoptosis in different cell types. The EMBO journal. 1996;15(16):4130-41. Epub 1996/08/15.
[96] Evan GI, Wyllie AH, Gilbert CS, Littlewood TD, Land H, Brooks M, et al. Induction of apoptosis in fibroblasts by c-myc protein. Cell. 1992;69(1):119-28. Epub 1992/04/03.
[97] Oda E, Ohki R, Murasawa H, Nemoto J, Shibue T, Yamashita T, et al. Noxa, a BH3-only member of the Bcl-2 family and candidate mediator of p53-induced apoptosis. Science. 2000;288(5468):1053-8. Epub 2000/05/12.
[98] Nakano K, Vousden KH. PUMA, a novel proapoptotic gene, is induced by p53. Molecular cell. 2001;7(3):683-94. Epub 2001/07/21.
[99] Eskes R, Desagher S, Antonsson B, Martinou JC. Bid induces the oligomerization and insertion of Bax into the outer mitochondrial membrane. Mol Cell Biol. 2000;20(3):929-35. Epub 2000/01/11.
[100] Korsmeyer SJ, Wei MC, Saito M, Weiler S, Oh KJ, Schlesinger PH. Pro-apoptotic cascade activates BID, which oligomerizes BAK or BAX into pores that result in the release of cytochrome c. Cell death and differentiation. 2000;7(12):1166-73. Epub 2001/02/15.
[101] Wei MC, Zong WX, Cheng EH, Lindsten T, Panoutsakopoulou V, Ross AJ, et al. Proapoptotic BAX and BAK: a requisite gateway to mitochondrial dysfunction and death. Science. 2001;292(5517):727-30. Epub 2001/04/28.
[102] Zong WX, Lindsten T, Ross AJ, MacGregor GR, Thompson CB. BH3-only proteins that bind pro-survival Bcl-2 family members fail to induce apoptosis in the absence of Bax and Bak. Genes & development. 2001;15(12):1481-6. Epub 2001/06/19.
[103] Wei MC, Lindsten T, Mootha VK, Weiler S, Gross A, Ashiya M, et al. tBID, a membrane-targeted death ligand, oligomerizes BAK to release cytochrome c. Genes & development. 2000;14(16):2060-71. Epub 2000/08/19.
[104] Kuwana T, Mackey MR, Perkins G, Ellisman MH, Latterich M, Schneiter R, et al. Bid, Bax, and lipids cooperate to form supramolecular openings in the outer mitochondrial membrane. Cell. 2002;111(3):331-42. Epub 2002/11/07.

[105] Kim H, Rafiuddin-Shah M, Tu HC, Jeffers JR, Zambetti GP, Hsieh JJ, et al. Hierarchical regulation of mitochondrion-dependent apoptosis by BCL-2 subfamilies. Nature cell biology. 2006;8(12):1348-58. Epub 2006/11/23.

[106] Sharpe JC, Arnoult D, Youle RJ. Control of mitochondrial permeability by Bcl-2 family members. Biochim Biophys Acta. 2004;1644(2-3):107-13. Epub 2004/03/05.

[107] Certo M, Del Gaizo Moore V, Nishino M, Wei G, Korsmeyer S, Armstrong SA, et al. Mitochondria primed by death signals determine cellular addiction to antiapoptotic BCL-2 family members. Cancer cell. 2006;9(5):351-65. Epub 2006/05/16.

[108] Cheng EH, Wei MC, Weiler S, Flavell RA, Mak TW, Lindsten T, et al. BCL-2, BCL-X(L) sequester BH3 domain-only molecules preventing BAX- and BAK-mediated mitochondrial apoptosis. Molecular cell. 2001;8(3):705-11. Epub 2001/10/05.

[109] Kuwana T, Bouchier-Hayes L, Chipuk JE, Bonzon C, Sullivan BA, Green DR, et al. BH3 domains of BH3-only proteins differentially regulate Bax-mediated mitochondrial membrane permeabilization both directly and indirectly. Molecular cell. 2005;17(4):525-35. Epub 2005/02/22.

[110] Letai A, Sorcinelli MD, Beard C, Korsmeyer SJ. Antiapoptotic BCL-2 is required for maintenance of a model leukemia. Cancer cell. 2004;6(3):241-9. Epub 2004/09/24.

[111] Willis SN, Chen L, Dewson G, Wei A, Naik E, Fletcher JI, et al. Proapoptotic Bak is sequestered by Mcl-1 and Bcl-xL, but not Bcl-2, until displaced by BH3-only proteins. Genes & development. 2005;19(11):1294-305. Epub 2005/05/20.

[112] Willis SN, Fletcher JI, Kaufmann T, van Delft MF, Chen L, Czabotar PE, et al. Apoptosis initiated when BH3 ligands engage multiple Bcl-2 homologs, not Bax or Bak. Science. 2007;315(5813):856-9. Epub 2007/02/10.

[113] Hsu YT, Wolter KG, Youle RJ. Cytosol-to-membrane redistribution of Bax and Bcl-X(L) during apoptosis. Proceedings of the National Academy of Sciences of the United States of America. 1997;94(8):3668-72. Epub 1997/04/15.

[114] Hsu YT, Youle RJ. Nonionic detergents induce dimerization among members of the Bcl-2 family. The Journal of biological chemistry. 1997;272(21):13829-34. Epub 1997/05/23.

[115] Dlugosz PJ, Billen LP, Annis MG, Zhu W, Zhang Z, Lin J, et al. Bcl-2 changes conformation to inhibit Bax oligomerization. The EMBO journal. 2006;25(11):2287-96. Epub 2006/04/28.

[116] Letai A, Bassik MC, Walensky LD, Sorcinelli MD, Weiler S, Korsmeyer SJ. Distinct BH3 domains either sensitize or activate mitochondrial apoptosis, serving as prototype cancer therapeutics. Cancer cell. 2002;2(3):183-92. Epub 2002/09/21.

[117] Deng J, Carlson N, Takeyama K, Dal Cin P, Shipp M, Letai A. BH3 profiling identifies three distinct classes of apoptotic blocks to predict response to ABT-737 and conventional chemotherapeutic agents. Cancer cell. 2007;12(2):171-85. Epub 2007/08/19.

[118] Tsujimoto Y, Cossman J, Jaffe E, Croce CM. Involvement of the bcl-2 gene in human follicular lymphoma. Science. 1985;228(4706):1440-3. Epub 1985/06/21.

[119] Hanada M, Delia D, Aiello A, Stadtmauer E, Reed JC. bcl-2 gene hypomethylation and high-level expression in B-cell chronic lymphocytic leukemia. Blood. 1993;82(6):1820-8. Epub 1993/09/15.

[120] Yip KW, Reed JC. Bcl-2 family proteins and cancer. Oncogene. 2008;27(50):6398-406. Epub 2008/10/29.

[121] Kelly PN, Strasser A. The role of Bcl-2 and its pro-survival relatives in tumourigenesis and cancer therapy. Cell death and differentiation. 2011;18(9):1414-24. Epub 2011/03/19.

[122] Krajewska M, Krajewski S, Epstein JI, Shabaik A, Sauvageot J, Song K, et al. Immunohistochemical analysis of bcl-2, bax, bcl-X, and mcl-1 expression in prostate cancers. The American journal of pathology. 1996;148(5):1567-76. Epub 1996/05/01.

[123] Konopleva M, Zhao S, Hu W, Jiang S, Snell V, Weidner D, et al. The anti-apoptotic genes Bcl-X(L) and Bcl-2 are over-expressed and contribute to chemoresistance of non-proliferating leukaemic CD34+ cells. British journal of haematology. 2002;118(2):521-34. Epub 2002/07/26.

[124] Reed JC. Bcl-2 family proteins: strategies for overcoming chemoresistance in cancer. Adv Pharmacol. 1997;41:501-32. Epub 1997/01/01.

[125] Rampino N, Yamamoto H, Ionov Y, Li Y, Sawai H, Reed JC, et al. Somatic frameshift mutations in the BAX gene in colon cancers of the microsatellite mutator phenotype. Science. 1997;275(5302):967-9. Epub 1997/02/14.

[126] Zhang L, Yu J, Park BH, Kinzler KW, Vogelstein B. Role of BAX in the apoptotic response to anticancer agents. Science. 2000;290(5493):989-92. Epub 2000/11/04.

[127] Garrison SP, Jeffers JR, Yang C, Nilsson JA, Hall MA, Rehg JE, et al. Selection against PUMA gene expression in Myc-driven B-cell lymphomagenesis. Mol Cell Biol. 2008;28(17):5391-402. Epub 2008/06/25.

[128] Karst AM, Dai DL, Martinka M, Li G. PUMA expression is significantly reduced in human cutaneous melanomas. Oncogene. 2005;24(6):1111-6. Epub 2005/02/04.

[129] Miyashita T, Krajewski S, Krajewska M, Wang HG, Lin HK, Liebermann DA, et al. Tumor suppressor p53 is a regulator of bcl-2 and bax gene expression in vitro and in vivo. Oncogene. 1994;9(6):1799-805. Epub 1994/06/01.

[130] Miyashita T, Reed JC. Tumor suppressor p53 is a direct transcriptional activator of the human bax gene. Cell. 1995;80(2):293-9. Epub 1995/01/27.

[131] Sax JK, Fei P, Murphy ME, Bernhard E, Korsmeyer SJ, El-Deiry WS. BID regulation by p53 contributes to chemosensitivity. Nature cell biology. 2002;4(11):842-9. Epub 2002/10/29.

[132] Yu J, Wang Z, Kinzler KW, Vogelstein B, Zhang L. PUMA mediates the apoptotic response to p53 in colorectal cancer cells. Proceedings of the National Academy of Sciences of the United States of America. 2003;100(4):1931-6. Epub 2003/02/08.

[133] Chipuk JE, Kuwana T, Bouchier-Hayes L, Droin NM, Newmeyer DD, Schuler M, et al. Direct activation of Bax by p53 mediates mitochondrial membrane permeabilization and apoptosis. Science. 2004;303(5660):1010-4. Epub 2004/02/14.

[134] Deng X, Gao F, Flagg T, Anderson J, May WS. Bcl2's flexible loop domain regulates p53 binding and survival. Mol Cell Biol. 2006;26(12):4421-34. Epub 2006/06/02.
[135] Shangary S, Johnson DE. Recent advances in the development of anticancer agents targeting cell death inhibitors in the Bcl-2 protein family. Leukemia : official journal of the Leukemia Society of America, Leukemia Research Fund, UK. 2003;17(8):1470-81. Epub 2003/07/30.
[136] Wang JL, Liu D, Zhang ZJ, Shan S, Han X, Srinivasula SM, et al. Structure-based discovery of an organic compound that binds Bcl-2 protein and induces apoptosis of tumor cells. Proceedings of the National Academy of Sciences of the United States of America. 2000;97(13):7124-9. Epub 2000/06/22.
[137] Wang S, Yang D, Lippman ME. Targeting Bcl-2 and Bcl-XL with nonpeptidic small-molecule antagonists. Seminars in oncology. 2003;30(5 Suppl 16):133-42. Epub 2003/11/13.
[138] Manero F, Gautier F, Gallenne T, Cauquil N, Gree D, Cartron PF, et al. The small organic compound HA14-1 prevents Bcl-2 interaction with Bax to sensitize malignant glioma cells to induction of cell death. Cancer research. 2006;66(5):2757-64. Epub 2006/03/03.
[139] An J, Chen Y, Huang Z. Critical upstream signals of cytochrome C release induced by a novel Bcl-2 inhibitor. The Journal of biological chemistry. 2004;279(18):19133-40. Epub 2004/02/18.
[140] Nguyen M, Marcellus RC, Roulston A, Watson M, Serfass L, Murthy Madiraju SR, et al. Small molecule obatoclax (GX15-070) antagonizes MCL-1 and overcomes MCL-1-mediated resistance to apoptosis. Proceedings of the National Academy of Sciences of the United States of America. 2007;104(49):19512-7. Epub 2007/11/28.
[141] Oltersdorf T, Elmore SW, Shoemaker AR, Armstrong RC, Augeri DJ, Belli BA, et al. An inhibitor of Bcl-2 family proteins induces regression of solid tumours. Nature. 2005;435(7042):677-81. Epub 2005/05/20.
[142] Walensky LD. From mitochondrial biology to magic bullet: navitoclax disarms BCL-2 in chronic lymphocytic leukemia. J Clin Oncol. 2012;30(5):554-7. Epub 2011/12/21.
[143] Roberts AW, Seymour JF, Brown JR, Wierda WG, Kipps TJ, Khaw SL, et al. Substantial susceptibility of chronic lymphocytic leukemia to BCL2 inhibition: results of a phase I study of navitoclax in patients with relapsed or refractory disease. J Clin Oncol. 2012;30(5):488-96. Epub 2011/12/21.
[144] Gandhi L, Camidge DR, Ribeiro de Oliveira M, Bonomi P, Gandara D, Khaira D, et al. Phase I study of Navitoclax (ABT-263), a novel Bcl-2 family inhibitor, in patients with small-cell lung cancer and other solid tumors. J Clin Oncol. 2011;29(7):909-16. Epub 2011/02/02.
[145] Meng Y, Tang W, Dai Y, Wu X, Liu M, Ji Q, et al. Natural BH3 mimetic (-)-gossypol chemosensitizes human prostate cancer via Bcl-xL inhibition accompanied by increase of Puma and Noxa. Molecular cancer therapeutics. 2008;7(7):2192-202. Epub 2008/07/23.

[146] Kitada S, Kress CL, Krajewska M, Jia L, Pellecchia M, Reed JC. Bcl-2 antagonist apogossypol (NSC736630) displays single-agent activity in Bcl-2-transgenic mice and has superior efficacy with less toxicity compared with gossypol (NSC19048). Blood. 2008;111(6):3211-9. Epub 2008/01/19.
[147] Sun Y, Wu J, Aboukameel A, Banerjee S, Arnold AA, Chen J, et al. Apogossypolone, a nonpeptidic small molecule inhibitor targeting Bcl-2 family proteins, effectively inhibits growth of diffuse large cell lymphoma cells in vitro and in vivo. Cancer Biol Ther. 2008;7(9):1418-26. Epub 2008/09/05.
[148] Al-Katib AM, Sun Y, Goustin AS, Azmi AS, Chen B, Aboukameel A, et al. SMI of Bcl-2 TW-37 is active across a spectrum of B-cell tumors irrespective of their proliferative and differentiation status. Journal of hematology & oncology. 2009;2:8. Epub 2009/02/18.
[149] Tamm I. AEG-35156, an antisense oligonucleotide against X-linked inhibitor of apoptosis for the potential treatment of cancer. Current opinion in investigational drugs. 2008;9(6):638-46. Epub 2008/06/03.
[150] Shimizu S, Kanaseki T, Mizushima N, Mizuta T, Arakawa-Kobayashi S, Thompson CB, et al. Role of Bcl-2 family proteins in a non-apoptotic programmed cell death dependent on autophagy genes. Nature cell biology. 2004;6(12):1221-8. Epub 2004/11/24.
[151] Oberstein A, Jeffrey PD, Shi Y. Crystal structure of the Bcl-XL-Beclin 1 peptide complex: Beclin 1 is a novel BH3-only protein. The Journal of biological chemistry. 2007;282(17):13123-32. Epub 2007/03/06.
[152] Pattingre S, Tassa A, Qu X, Garuti R, Liang XH, Mizushima N, et al. Bcl-2 antiapoptotic proteins inhibit Beclin 1-dependent autophagy. Cell. 2005;122(6):927-39. Epub 2005/09/24.
[153] Chen S, Rehman S, Zhang W, Wen A, Yao L, Zhang J. Autophagy is a therapeutic target in anticancer drug resistance. Biochimica et biophysica acta. 2010;1806(c2ea68e4-65ce-52d4-86d3-fb085eb006dc):220-9.
[154] Mizushima N, Levine B, Cuervo AM, Klionsky DJ. Autophagy fights disease through cellular self-digestion. Nature. 2008;451(7182):1069-75. Epub 2008/02/29.
[155] Lucocq J, Walker D. Evidence for fusion between multilamellar endosomes and autophagosomes in HeLa cells. European journal of cell biology. 1997;72(4):307-13. Epub 1997/04/01.
[156] Vazquez-Martin A, Oliveras-Ferraros C, Menendez J. Autophagy facilitates the development of breast cancer resistance to the anti-HER2 monoclonal antibody trastuzumab. PloS one. 2009;4(1fdb94e2-03b9-d18b-5b7d-9bf165abe7df).
[157] Abedin MJ, Wang D, McDonnell MA, Lehmann U, Kelekar A. Autophagy delays apoptotic death in breast cancer cells following DNA damage. Cell death and differentiation. 2007;14(3):500-10. Epub 2006/09/23.
[158] Liu D, Yang Y, Liu Q, Wang J. Inhibition of autophagy by 3-MA potentiates cisplatin-induced apoptosis in esophageal squamous cell carcinoma cells. Med Oncol. 2011;28(1):105-11. Epub 2009/12/31.

[159] Ren JH, He WS, Nong L, Zhu QY, Hu K, Zhang RG, et al. Acquired cisplatin resistance in human lung adenocarcinoma cells is associated with enhanced autophagy. Cancer biotherapy & radiopharmaceuticals. 2010;25(1):75-80. Epub 2010/03/02.

[160] Djavaheri-Mergny M, Amelotti M, Mathieu J, Besancon F, Bauvy C, Codogno P. Regulation of autophagy by NFkappaB transcription factor and reactives oxygen species. Autophagy. 2007;3(4):390-2. Epub 2007/05/02.

[161] Itakura E, Mizushima N. Atg14 and UVRAG: mutually exclusive subunits of mammalian Beclin 1-PI3K complexes. Autophagy. 2009;5(4):534-6. Epub 2009/02/19.

[162] Trocoli A, Djavaheri-Mergny M. The complex interplay between autophagy and NF-kappaB signaling pathways in cancer cells. Am J Cancer Res. 2011;1(5):629-49. Epub 2011/10/14.

[163] Shen S, Kepp O, Kroemer G. The end of autophagic cell death? Autophagy. 2012;8(1):1-3. Epub 2011/11/16.

[164] Yue Z, Jin S, Yang C, Levine AJ, Heintz N. Beclin 1, an autophagy gene essential for early embryonic development, is a haploinsufficient tumor suppressor. Proceedings of the National Academy of Sciences of the United States of America. 2003;100(25):15077-82. Epub 2003/12/06.

[165] Qu X, Yu J, Bhagat G, Furuya N, Hibshoosh H, Troxel A, et al. Promotion of tumorigenesis by heterozygous disruption of the beclin 1 autophagy gene. The Journal of clinical investigation. 2003;112(12):1809-20. Epub 2003/11/26.

[166] Jung CH, Jun CB, Ro SH, Kim YM, Otto NM, Cao J, et al. ULK-Atg13-FIP200 complexes mediate mTOR signaling to the autophagy machinery. Mol Biol Cell. 2009;20(7):1992-2003. Epub 2009/02/20.

[167] Hosokawa N, Hara T, Kaizuka T, Kishi C, Takamura A, Miura Y, et al. Nutrient-dependent mTORC1 association with the ULK1-Atg13-FIP200 complex required for autophagy. Molecular biology of the cell. 2009;20(121ccf76-f86a-2e4b-8ca9-af545e417259):1981-2072.

[168] Hara T, Takamura A, Kishi C, Iemura S, Natsume T, Guan JL, et al. FIP200, a ULK-interacting protein, is required for autophagosome formation in mammalian cells. J Cell Biol. 2008;181(3):497-510. Epub 2008/04/30.

[169] He C, Levine B. The Beclin 1 interactome. Current opinion in cell biology. 2010;22(2):140-9. Epub 2010/01/26.

[170] Obara K, Ohsumi Y. Atg14: a key player in orchestrating autophagy. Int J Cell Biol. 2011;2011:713435. Epub 2011/10/21.

[171] Axe EL, Walker SA, Manifava M, Chandra P, Roderick HL, Habermann A, et al. Autophagosome formation from membrane compartments enriched in phosphatidylinositol 3-phosphate and dynamically connected to the endoplasmic reticulum. J Cell Biol. 2008;182(4):685-701. Epub 2008/08/30.

[172] Fan W, Nassiri A, Zhong Q. Autophagosome targeting and membrane curvature sensing by Barkor/Atg14(L). Proceedings of the National Academy of Sciences of the United States of America. 2011;108(19):7769-74. Epub 2011/04/27.

[173] Mizushima N, Sugita H, Yoshimori T, Ohsumi Y. A new protein conjugation system in human. The counterpart of the yeast Apg12p conjugation system essential for autophagy. J Biol Chem. 1998;273(51):33889-92. Epub 1998/12/16.

[174] Mizushima N, Yoshimori T, Ohsumi Y. Mouse Apg10 as an Apg12-conjugating enzyme: analysis by the conjugation-mediated yeast two-hybrid method. FEBS Lett. 2002;532(3):450-4. Epub 2002/12/17.

[175] Tanida I, Tanida-Miyake E, Ueno T, Kominami E. The human homolog of Saccharomyces cerevisiae Apg7p is a Protein-activating enzyme for multiple substrates including human Apg12p, GATE-16, GABARAP, and MAP-LC3. J Biol Chem. 2001;276(3):1701-6. Epub 2000/11/30.

[176] Mizushima N, Kuma A, Kobayashi Y, Yamamoto A, Matsubae M, Takao T, et al. Mouse Apg16L, a novel WD-repeat protein, targets to the autophagic isolation membrane with the Apg12-Apg5 conjugate. J Cell Sci. 2003;116(Pt 9):1679-88. Epub 2003/04/01.

[177] Ichimura Y, Kirisako T, Takao T, Satomi Y, Shimonishi Y, Ishihara N, et al. A ubiquitin-like system mediates protein lipidation. Nature. 2000;408(6811):488-92. Epub 2000/12/02.

[178] Kirisako T, Ichimura Y, Okada H, Kabeya Y, Mizushima N, Yoshimori T, et al. The reversible modification regulates the membrane-binding state of Apg8/Aut7 essential for autophagy and the cytoplasm to vacuole targeting pathway. The Journal of cell biology. 2000;151(2):263-76. Epub 2000/10/19.

[179] Fujita N, Itoh T, Omori H, Fukuda M, Noda T, Yoshimori T. The Atg16L complex specifies the site of LC3 lipidation for membrane biogenesis in autophagy. Mol Biol Cell. 2008;19(5):2092-100. Epub 2008/03/07.

[180] Jager S, Bucci C, Tanida I, Ueno T, Kominami E, Saftig P, et al. Role for Rab7 in maturation of late autophagic vacuoles. J Cell Sci. 2004;117(Pt 20):4837-48. Epub 2004/09/02.

[181] Kimura S, Noda T, Yoshimori T. Dissection of the autophagosome maturation process by a novel reporter protein, tandem fluorescent-tagged LC3. Autophagy. 2007;3(5):452-60. Epub 2007/05/31.

[182] Nobukuni T, Joaquin M, Roccio M, Dann SG, Kim SY, Gulati P, et al. Amino acids mediate mTOR/raptor signaling through activation of class 3 phosphatidylinositol 3OH-kinase. Proceedings of the National Academy of Sciences of the United States of America. 2005;102(40):14238-43. Epub 2005/09/24.

[183] Lum JJ, Bauer DE, Kong M, Harris MH, Li C, Lindsten T, et al. Growth factor regulation of autophagy and cell survival in the absence of apoptosis. Cell. 2005;120(2):237-48. Epub 2005/02/01.

[184] Li Y, Corradetti MN, Inoki K, Guan KL. TSC2: filling the GAP in the mTOR signaling pathway. Trends in biochemical sciences. 2004;29(1):32-8. Epub 2004/01/20.

[185] Li J, Yen C, Liaw D, Podsypanina K, Bose S, Wang SI, et al. PTEN, a putative protein tyrosine phosphatase gene mutated in human brain, breast, and prostate cancer. Science. 1997;275(5308):1943-7. Epub 1997/03/28.

[186] Stommel JM, Kimmelman AC, Ying H, Nabioullin R, Ponugoti AH, Wiedemeyer R, et al. Coactivation of receptor tyrosine kinases affects the response of tumor cells to targeted therapies. Science. 2007;318(5848):287-90. Epub 2007/09/18.

[187] Chen CT, Kim H, Liska D, Gao S, Christensen JG, Weiser MR. MET activation mediates resistance to lapatinib inhibition of HER2-amplified gastric cancer cells. Molecular cancer therapeutics. 2012;11(3):660-9. Epub 2012/01/13.

[188] Gwinn DM, Shackelford DB, Egan DF, Mihaylova MM, Mery A, Vasquez DS, et al. AMPK phosphorylation of raptor mediates a metabolic checkpoint. Molecular cell. 2008;30(2):214-26. Epub 2008/04/29.

[189] Papandreou I, Lim AL, Laderoute K, Denko NC. Hypoxia signals autophagy in tumor cells via AMPK activity, independent of HIF-1, BNIP3, and BNIP3L. Cell death and differentiation. 2008;15(10):1572-81. Epub 2008/06/14.

[190] Hoyer-Hansen M, Jaattela M. AMP-activated protein kinase: a universal regulator of autophagy? Autophagy. 2007;3(4):381-3. Epub 2007/04/26.

[191] Kim J, Kundu M, Viollet B, Guan KL. AMPK and mTOR regulate autophagy through direct phosphorylation of Ulk1. Nature cell biology. 2011;13(2):132-41. Epub 2011/01/25.

[192] Crighton D, Wilkinson S, O'Prey J, Syed N, Smith P, Harrison PR, et al. DRAM, a p53-induced modulator of autophagy, is critical for apoptosis. Cell. 2006;126(1):121-34. Epub 2006/07/15.

[193] Budanov AV, Karin M. p53 target genes sestrin1 and sestrin2 connect genotoxic stress and mTOR signaling. Cell. 2008;134(3):451-60. Epub 2008/08/12.

[194] Gao W, Shen Z, Shang L, Wang X. Upregulation of human autophagy-initiation kinase ULK1 by tumor suppressor p53 contributes to DNA-damage-induced cell death. Cell death and differentiation. 2011;18(10):1598-607. Epub 2011/04/09.

[195] Carter BZ, Mak DH, Morris SJ, Borthakur G, Estey E, Byrd AL, et al. XIAP antisense oligonucleotide (AEG35156) achieves target knockdown and induces apoptosis preferentially in CD34+38- cells in a phase 1/2 study of patients with relapsed/refractory AML. Apoptosis : an international journal on programmed cell death. 2011;16(1):67-74. Epub 2010/10/13.

[196] Rom J, von Minckwitz G, Eiermann W, Sievert M, Schlehe B, Marme F, et al. Oblimersen combined with docetaxel, adriamycin and cyclophosphamide as neo-adjuvant systemic treatment in primary breast cancer: final results of a multicentric phase I study. Annals of oncology : official journal of the European Society for Medical Oncology / ESMO. 2008;19(10):1698-705. Epub 2008/05/15.

[197] Liang XH, Jackson S, Seaman M, Brown K, Kempkes B, Hibshoosh H, et al. Induction of autophagy and inhibition of tumorigenesis by beclin 1. Nature. 1999;402(6762):672-6. Epub 1999/12/22.

[198] Miracco C, Cosci E, Oliveri G, Luzi P, Pacenti L, Monciatti I, et al. Protein and mRNA expression of autophagy gene Beclin 1 in human brain tumours. Int J Oncol. 2007;30(2):429-36. Epub 2007/01/05.

[199] Shi YH, Ding ZB, Zhou J, Qiu SJ, Fan J. Prognostic significance of Beclin 1-dependent apoptotic activity in hepatocellular carcinoma. Autophagy. 2009;5(3):380-2. Epub 2009/01/16.

[200] Ahn CH, Jeong EG, Lee JW, Kim MS, Kim SH, Kim SS, et al. Expression of beclin-1, an autophagy-related protein, in gastric and colorectal cancers. APMIS : acta pathologica, microbiologica, et immunologica Scandinavica. 2007;115(12):1344-9. Epub 2008/01/11.

[201] Akar U, Chaves-Reyez A, Barria M, Tari A, Sanguino A, Kondo Y, et al. Silencing of Bcl-2 expression by small interfering RNA induces autophagic cell death in MCF-7 breast cancer cells. Autophagy. 2008;4(5):669-79. Epub 2008/04/22.

[202] Di Bartolomeo S, Corazzari M, Nazio F, Oliverio S, Lisi G, Antonioli M, et al. The dynamic interaction of AMBRA1 with the dynein motor complex regulates mammalian autophagy. The Journal of cell biology. 2010;191(1):155-68. Epub 2010/10/06.

[203] Strappazzon F, Vietri-Rudan M, Campello S, Nazio F, Florenzano F, Fimia GM, et al. Mitochondrial BCL-2 inhibits AMBRA1-induced autophagy. The EMBO journal. 2011;30(7):1195-208. Epub 2011/03/02.

[204] Tang D, Kang R, Cheh CW, Livesey KM, Liang X, Schapiro NE, et al. HMGB1 release and redox regulates autophagy and apoptosis in cancer cells. Oncogene. 2010;29(38):5299-310. Epub 2010/07/14.

[205] Kang R, Livesey K, Zeh H, Loze M, Tang D. HMGB1: a novel Beclin 1-binding protein active in autophagy. Autophagy. 2010;6(99a93e1e-c67d-83e4-87ca-9aea803ddcd1):1209-20.

[206] Janku F, McConkey D, Hong D, Kurzrock R. Autophagy as a target for anticancer therapy. Nature reviews Clinical oncology. 2011;8(ee0c5394-e2da-32d4-597e-9b23fbfbab39):528-67.

[207] Wirawan E, Vande Walle L, Kersse K, Cornelis S, Claerhout S, Vanoverberghe I, et al. Caspase-mediated cleavage of Beclin-1 inactivates Beclin-1-induced autophagy and enhances apoptosis by promoting the release of proapoptotic factors from mitochondria. Cell death & disease. 2010;1:e18. Epub 2010/01/01.

[208] Maiuri M, Criollo A, Tasdemir E, Vicencio J, Tajeddine N, Hickman J, et al. BH3-only proteins and BH3 mimetics induce autophagy by competitively disrupting the interaction between Beclin 1 and Bcl-2/Bcl-X(L). Autophagy.3(f1536ce7-df32-56eb-f51e-7b9f6f70c571):374-80.

[209] Satoh T, Okamoto I, Miyazaki M, Morinaga R, Tsuya A, Hasegawa Y, et al. Phase I study of YM155, a novel survivin suppressant, in patients with advanced solid tumors. Clinical cancer research : an official journal of the American Association for Cancer Research. 2009;15(11):3872-80. Epub 2009/05/28.

[210] Giaccone G, Zatloukal P, Roubec J, Floor K, Musil J, Kuta M, et al. Multicenter phase II trial of YM155, a small-molecule suppressor of survivin, in patients with advanced, refractory, non-small-cell lung cancer. J Clin Oncol. 2009;27(27):4481-6. Epub 2009/08/19.

[211] Malik SA, Orhon I, Morselli E, Criollo A, Shen S, Marino G, et al. BH3 mimetics activate multiple pro-autophagic pathways. Oncogene. 2011;30(37):3918-29. Epub 2011/04/05.

[212] Han W, Pan H, Chen Y, Sun J, Wang Y, Li J, et al. EGFR tyrosine kinase inhibitors activate autophagy as a cytoprotective response in human lung cancer cells. PloS one. 2011;6(79e9b519-257f-e019-7009-9b903ae72bc4).

[213] Swampillai AL, Salomoni P, Short SC. The Role of Autophagy in Clinical Practice. Clin Oncol (R Coll Radiol). 2011. Epub 2011/10/29.

[214] Huang J, Ni J, Liu K, Yu Y, Xie M, Kang R, et al. HMGB1 promotes drug resistance in osteosarcoma. Cancer research. 2012;72(f06c5ead-ce77-b4c8-b289-95ddcffe93ca):230-8.

[215] Chaachouay H, Ohneseit P, Toulany M, Kehlbach R, Multhoff G, Rodemann H. Autophagy contributes to resistance of tumor cells to ionizing radiation. Radiotherapy and oncology : journal of the European Society for Therapeutic Radiology and Oncology. 2011;99(f586ccb5-12b0-be73-4db2-fb085eb012a7):287-379.

[216] Blommaart EF, Krause U, Schellens JP, Vreeling-Sindelarova H, Meijer AJ. The phosphatidylinositol 3-kinase inhibitors wortmannin and LY294002 inhibit autophagy in isolated rat hepatocytes. European journal of biochemistry / FEBS. 1997;243(1-2):240-6. Epub 1997/01/15.

[217] Wu Y-T, Tan H-L, Shui G, Bauvy C, Huang Q, Wenk M, et al. Dual role of 3-methyladenine in modulation of autophagy via different temporal patterns of inhibition on class I and III phosphoinositide 3-kinase. The Journal of biological chemistry. 2010;285(69711380-4158-c2c6-9a36-95f349333c77):10850-911.

[218] Bellodi C, Lidonnici MR, Hamilton A, Helgason GV, Soliera AR, Ronchetti M, et al. Targeting autophagy potentiates tyrosine kinase inhibitor-induced cell death in Philadelphia chromosome-positive cells, including primary CML stem cells. The Journal of clinical investigation. 2009;119(5):1109-23. Epub 2009/04/14.

[219] Shen HM, Codogno P. Autophagic cell death: Loch Ness monster or endangered species? Autophagy. 2011;7(5):457-65. Epub 2010/12/15.

[220] Solomon VR, Lee H. Chloroquine and its analogs: a new promise of an old drug for effective and safe cancer therapies. European journal of pharmacology. 2009;625(1-3):220-33. Epub 2009/10/20.

[221] Swampillai A, Salomoni P, Short S. The Role of Autophagy in Clinical Practice. Clinical oncology (Royal College of Radiologists (Great Britain)). 2011(02458d04-8bf9-caed-2448-fb085eba55e5).

[222] Thorburn J, Frankel A, Thorburn A. Regulation of HMGB1 release by autophagy. Autophagy. 2009;5(0c807524-d29e-a4fe-3f06-9a0cae660f50):247-56.

[223] Apetoh L, Ghiringhelli F, Tesniere A, Criollo A, Ortiz C, Lidereau R, et al. The interaction between HMGB1 and TLR4 dictates the outcome of anticancer

chemotherapy and radiotherapy. Immunological reviews. 2007;220:47-59. Epub 2007/11/06.

[224] Thorburn J, Horita H, Redzic J, Hansen K, Frankel AE, Thorburn A. Autophagy regulates selective HMGB1 release in tumor cells that are destined to die. Cell death and differentiation. 2009;16(1):175-83. Epub 2008/10/11.

[225] Cai Q, Sun H, Peng Y, Lu J, Nikolovska-Coleska Z, McEachern D, et al. A potent and orally active antagonist (SM-406/AT-406) of multiple inhibitor of apoptosis proteins (IAPs) in clinical development for cancer treatment. Journal of medicinal chemistry. 2011;54(8):2714-26. Epub 2011/03/30.

[226] Baggstrom MQ, Qi Y, Koczywas M, Argiris A, Johnson EA, Millward MJ, et al. A phase II study of AT-101 (Gossypol) in chemotherapy-sensitive recurrent extensive-stage small cell lung cancer. Journal of thoracic oncology : official publication of the International Association for the Study of Lung Cancer. 2011;6(10):1757-60. Epub 2011/09/16.

[227] Ready N, Karaseva NA, Orlov SV, Luft AV, Popovych O, Holmlund JT, et al. Double-blind, placebo-controlled, randomized phase 2 study of the proapoptotic agent AT-101 plus docetaxel, in second-line non-small cell lung cancer. Journal of thoracic oncology : official publication of the International Association for the Study of Lung Cancer. 2011;6(4):781-5. Epub 2011/02/04.

[228] Heist RS, Fain J, Chinnasami B, Khan W, Molina JR, Sequist LV, et al. Phase I/II study of AT-101 with topotecan in relapsed and refractory small cell lung cancer. Journal of thoracic oncology : official publication of the International Association for the Study of Lung Cancer. 2010;5(10):1637-43. Epub 2010/09/03.

[229] Liu G, Kelly WK, Wilding G, Leopold L, Brill K, Somer B. An open-label, multicenter, phase I/II study of single-agent AT-101 in men with castrate-resistant prostate cancer. Clinical cancer research : an official journal of the American Association for Cancer Research. 2009;15(9):3172-6. Epub 2009/04/16.

[230] Van Poznak C, Seidman AD, Reidenberg MM, Moasser MM, Sklarin N, Van Zee K, et al. Oral gossypol in the treatment of patients with refractory metastatic breast cancer: a phase I/II clinical trial. Breast cancer research and treatment. 2001;66(3):239-48. Epub 2001/08/21.

[231] Wilson WH, O'Connor OA, Czuczman MS, LaCasce AS, Gerecitano JF, Leonard JP, et al. Navitoclax, a targeted high-affinity inhibitor of BCL-2, in lymphoid malignancies: a phase 1 dose-escalation study of safety, pharmacokinetics, pharmacodynamics, and antitumour activity. The lancet oncology. 2010;11(12):1149-59. Epub 2010/11/26.

[232] Schimmer AD, O'Brien S, Kantarjian H, Brandwein J, Cheson BD, Minden MD, et al. A phase I study of the pan bcl-2 family inhibitor obatoclax mesylate in patients with advanced hematologic malignancies. Clinical cancer research : an official journal of the American Association for Cancer Research. 2008;14(24):8295-301. Epub 2008/12/18.

[233] Paik PK, Rudin CM, Brown A, Rizvi NA, Takebe N, Travis W, et al. A phase I study of obatoclax mesylate, a Bcl-2 antagonist, plus topotecan in solid tumor malignancies. Cancer chemotherapy and pharmacology. 2010;66(6):1079-85. Epub 2010/02/19.

[234] O'Brien SM, Claxton DF, Crump M, Faderl S, Kipps T, Keating MJ, et al. Phase I study of obatoclax mesylate (GX15-070), a small molecule pan-Bcl-2 family antagonist, in patients with advanced chronic lymphocytic leukemia. Blood. 2009;113(2):299-305. Epub 2008/10/22.

[235] Rom J, von Minckwitz G, Marme F, Ataseven B, Kozian D, Sievert M, et al. Phase I study of apoptosis gene modulation with oblimersen within preoperative chemotherapy in patients with primary breast cancer. Annals of oncology : official journal of the European Society for Medical Oncology / ESMO. 2009;20(11):1829-35. Epub 2009/07/17.

[236] Sternberg CN, Dumez H, Van Poppel H, Skoneczna I, Sella A, Daugaard G, et al. Docetaxel plus oblimersen sodium (Bcl-2 antisense oligonucleotide): an EORTC multicenter, randomized phase II study in patients with castration-resistant prostate cancer. Annals of oncology : official journal of the European Society for Medical Oncology / ESMO. 2009;20(7):1264-9. Epub 2009/03/20.

[237] Tolcher AW, Chi K, Kuhn J, Gleave M, Patnaik A, Takimoto C, et al. A phase II, pharmacokinetic, and biological correlative study of oblimersen sodium and docetaxel in patients with hormone-refractory prostate cancer. Clinical cancer research : an official journal of the American Association for Cancer Research. 2005;11(10):3854-61. Epub 2005/05/18.

Permissions

The contributors of this book come from diverse backgrounds, making this book a truly international effort. This book will bring forth new frontiers with its revolutionizing research information and detailed analysis of the nascent developments around the world.

We would like to thank Justine Rudner, for lending his expertise to make the book truly unique. He has played a crucial role in the development of this book. Without his invaluable contribution this book wouldn't have been possible. He has made vital efforts to compile up to date information on the varied aspects of this subject to make this book a valuable addition to the collection of many professionals and students.

This book was conceptualized with the vision of imparting up-to-date information and advanced data in this field. To ensure the same, a matchless editorial board was set up. Every individual on the board went through rigorous rounds of assessment to prove their worth. After which they invested a large part of their time researching and compiling the most relevant data for our readers. Conferences and sessions were held from time to time between the editorial board and the contributing authors to present the data in the most comprehensible form. The editorial team has worked tirelessly to provide valuable and valid information to help people across the globe.

Every chapter published in this book has been scrutinized by our experts. Their significance has been extensively debated. The topics covered herein carry significant findings which will fuel the growth of the discipline. They may even be implemented as practical applications or may be referred to as a beginning point for another development. Chapters in this book were first published by InTech; hereby published with permission under the Creative Commons Attribution License or equivalent.

The editorial board has been involved in producing this book since its inception. They have spent rigorous hours researching and exploring the diverse topics which have resulted in the successful publishing of this book. They have passed on their knowledge of decades through this book. To expedite this challenging task, the publisher supported the team at every step. A small team of assistant editors was also appointed to further simplify the editing procedure and attain best results for the readers.

Our editorial team has been hand-picked from every corner of the world. Their multi-ethnicity adds dynamic inputs to the discussions which result in innovative outcomes. These outcomes are then further discussed with the researchers and contributors who give their valuable feedback and opinion regarding the same. The feedback is then collaborated with the researches and they are edited in a comprehensive manner to aid the understanding of the subject.

Apart from the editorial board, the designing team has also invested a significant amount of their time in understanding the subject and creating the most relevant covers. They scrutinized every image to scout for the most suitable representation of the subject and create an appropriate cover for the book.

The publishing team has been involved in this book since its early stages. They were actively engaged in every process, be it collecting the data, connecting with the contributors or procuring relevant information. The team has been an ardent support to the editorial, designing and production team. Their endless efforts to recruit the best for this project, has resulted in the accomplishment of this book. They are a veteran in the field of academics and their pool of knowledge is as vast as their experience in printing. Their expertise and guidance has proved useful at every step. Their uncompromising quality standards have made this book an exceptional effort. Their encouragement from time to time has been an inspiration for everyone.

The publisher and the editorial board hope that this book will prove to be a valuable piece of knowledge for researchers, students, practitioners and scholars across the globe.

List of Contributors

Jamie C. Stanford and Rebecca S. Cook
Department of Cancer Biology, Vanderbilt University, Nashville, TN, USA

Joaquín H. Patarroyo S.
Laboratory of Biology and Control of Hematozoa and Vectors, BIOAGRO/Veterinary Department - Federal Viçosa University Viçosa-MG, Brazil

Marlene I. Vargas V.
Laboratory of Immunopathology, BIOAGRO/Veterinary Department - Federal Viçosa University Viçosa-MG, Brazil

Mona Desai and Qigui Yu
Center for AIDS Research, Department of Microbiology and Immunology, Indiana University School of Medicine, Indianapolis, Indiana, USA Division of Infectious Diseases, Department of Medicine, Indiana University School of Medicine, Indianapolis, Indiana, USA

Daniel Byrd
Center for AIDS Research, Department of Microbiology and Immunology, Indiana University School of Medicine, Indianapolis, Indiana, USA

Ningjie Hu
Zhejiang Provincial Key Laboratory for Technology & Application of Model Organisms, Wenzhou Medical College, University Park, Wenzhou, China

Carmen Sanges, Nunzia Migliaccio, Paolo Arcari and Annalisa Lamberti
Department of Biochemistry and Medical Biotechnologies, University of Naples Federico II, Naples, Italy CEINGE, Advanced Biotechnologies scarl, Naples, Italy

Masahiko Takemura
Department of Genetics, Cell Biology and Development, University of Minnesota, Minneapolis, Minnesota, USA Department of Life Science, Faculty of Science, Gakushuin University, 1-5-1 Mejiro, Toshima-ku, Tokyo, Japan

Takashi Adachi-Yamada
Department of Life Science, Faculty of Science, Gakushuin University, 1-5-1 Mejiro, Toshima-ku, Tokyo, Japan

Gregory Lucien Bellot
Department of Physiology, Yong Loo Lin School of Medicine, National University of Singapore, Singapore

Xueying Wang
Department of Biochemistry, Yong Loo Lin School of Medicine, National University of Singapore, Singapore

Qian Nancy Hu and Troy A. Baldwin
University of Alberta, Canada

Rebecca T. Marquez, Bryan W. Tsao and Nicholas F. Faust
Department of Molecular Biosciences, University of Kansas, Lawrence, Kansas, USA

Liang Xu
Department of Molecular Biosciences, University of Kansas, Lawrence, Kansas, USA Departments of Radiation Oncology and Urology, University of Kansas Medical School, Kansas City, Kansas, USA Departments of Radiation Oncology and Urology, University of Kansas Cancer Center, Kansas City, Kansas, USA

Printed in the USA
CPSIA information can be obtained
at www.ICGtesting.com
JSHW011402221024
72173JS00003B/390

9 781632 390745